Christian Zschalig

Characterizations of Planar Lattices

Christian Zschalig

Characterizations of Planar Lattices

An Approach based on Left-relations

Südwestdeutscher Verlag für Hochschulschriften

Impressum/Imprint (nur für Deutschland/ only for Germany)
Bibliografische Information der Deutschen Nationalbibliothek: Die Deutsche Nationalbibliothek verzeichnet diese Publikation in der Deutschen Nationalbibliografie; detaillierte bibliografische Daten sind im Internet über http://dnb.d-nb.de abrufbar.
Alle in diesem Buch genannten Marken und Produktnamen unterliegen warenzeichen-, marken- oder patentrechtlichem Schutz bzw. sind Warenzeichen oder eingetragene Warenzeichen der jeweiligen Inhaber. Die Wiedergabe von Marken, Produktnamen, Gebrauchsnamen, Handelsnamen, Warenbezeichnungen u.s.w. in diesem Werk berechtigt auch ohne besondere Kennzeichnung nicht zu der Annahme, dass solche Namen im Sinne der Warenzeichen- und Markenschutzgesetzgebung als frei zu betrachten wären und daher von jedermann benutzt werden dürften.

Verlag: Südwestdeutscher Verlag für Hochschulschriften Aktiengesellschaft & Co. KG
Dudweiler Landstr. 99, 66123 Saarbrücken, Deutschland
Telefon +49 681 37 20 271-1, Telefax +49 681 37 20 271-0, Email: info@svh-verlag.de
Zugl.: Dresden, TU, Diss., 2009

Herstellung in Deutschland:
Schaltungsdienst Lange o.H.G., Berlin
Books on Demand GmbH, Norderstedt
Reha GmbH, Saarbrücken
Amazon Distribution GmbH, Leipzig
ISBN: 978-3-8381-0808-7

Imprint (only for USA, GB)
Bibliographic information published by the Deutsche Nationalbibliothek: The Deutsche Nationalbibliothek lists this publication in the Deutsche Nationalbibliografie; detailed bibliographic data are available in the Internet at http://dnb.d-nb.de.
Any brand names and product names mentioned in this book are subject to trademark, brand or patent protection and are trademarks or registered trademarks of their respective holders. The use of brand names, product names, common names, trade names, product descriptions etc. even without a particular marking in this works is in no way to be construed to mean that such names may be regarded as unrestricted in respect of trademark and brand protection legislation and could thus be used by anyone.

Publisher:
Südwestdeutscher Verlag für Hochschulschriften Aktiengesellschaft & Co. KG
Dudweiler Landstr. 99, 66123 Saarbrücken, Germany
Phone +49 681 37 20 271-1, Fax +49 681 37 20 271-0, Email: info@svh-verlag.de

Copyright © 2009 by the author and Südwestdeutscher Verlag für Hochschulschriften Aktiengesellschaft & Co. KG and licensors
All rights reserved. Saarbrücken 2009

Printed in the U.S.A.
Printed in the U.K. by (see last page)
ISBN: 978-3-8381-0808-7

Danksagung

Ganz herzlich möchte ich mich bei allen bedanken, die mich – in welcher Art auch immer unterstützt haben, diese Dissertation schreiben zu können.

Zuallererst geht mein größter Dank an meine Freundin Susi Müller. Vor allem in der letzten, hektischen Phase der Dissertation hat sie mir die Kraft und Ruhe gegeben, die Arbeit vollenden zu können. Es ist wunderbar, dass jeder meiner Tage durch einen Sonnenstrahl erhellt wird.

Ein herzlicher Dank gebührt meinem Betreuer Bernhard Ganter; für die Freiheit, die er mir während der Promotionsphase ließ, ein Thema nach meinen Vorstellungen zu bearbeiten und für die Liebe zur Algebra, die er in mir geweckt hat.

Immer eine große Hilfe im universitären Alltag war mir Ulrike Baumann. Sie ermutigte mich, mit der Arbeit als Tutor zu beginnen und unterstützte mich in all den Jahren bei meiner Lehrtätigkeit. Ohne sie hätte ich mit der Promotion sicher nicht begonnen.

Meine ersten wissenschaftlichen Gehversuche unternahm ich meist zusammen mit Heiko Reppe. Vielen Dank ihm für das viele Korrekturlesen und die gemeinsamen Konferenzbesuche. Ich hoffe, ich konnte ihn in ähnlicher Weise unterstützen wie er mich.

Allen noch nicht namentlich genannten Mitgliedern des Instituts für Algebra sei für die warmherzige und unverkrampfte Atmosphäre gedankt, durch die Spaß und Freude an Forschung und Lehre in mir geweckt und erhalten wurden.

Vielen Dank an Jojo Klein, Phillip Olk, Kerstin Knott, Sanni Kehr, Ronny Peters, Ina Schaarschmidt und alle anderen von der „Bierstubencrew" für die vielen Kaffees und die entspannenden Gespräche, die ich mit euch genießen konnte.

Zum Schluß, aber keinesfalls am wenigsten möchte ich meinen Eltern Klaus und Maria Zschalig danken, die meine mathematischen Neigungen schon von kleinauf gefördert haben und die mir jederzeit jegliche Unterstützung zukommen ließen.

2

Contents

1	**Introduction**	**5**
2	**Motivation**	**6**
	2.1 Why Do We Need Nice Diagrams?	7
	2.2 What Is a Nice Diagram?	9
	2.3 Why Planarity?	10
3	**Preliminaries**	**12**
	3.1 Basic Definitions	13
	3.2 Diagrams of Ordered Structures	17
	3.3 Planar Lattices	21
	3.4 Plane Diagrams	27
	3.5 Formal Concept Analysis	30
	3.6 Ferrers-relations	32
4	**Left-relations on Lattices**	**35**
	4.1 Definition and Basic Properties	36
	4.2 Left-relations and Conjugate Orders	40
	4.3 Planarity Conditions	42
	4.4 Planar Contexts	48
5	**Left-relations on Diagrams**	**56**
	5.1 Definition	57
	5.2 Coherence to Left-relations on Lattices	60
	5.3 How to Draw Plane Diagrams	65
	5.3.1 Attribute Additive Diagrams	66
	5.3.2 Layer Diagrams	68
	5.3.3 Layered Attribute Additive Diagrams	70
6	**Left-relations on Contexts**	**71**
	6.1 Definition and First Properties	72
	6.2 The Components of Ferrers-graphs	77

		6.2.1	Connections in Intervals	79
		6.2.2	Induced Relations on Components of Γ	83
		6.2.3	Edge Sequences in Γ and Connections in $\mathfrak{V}$	90
		6.2.4	Components of Type 1 and Type 2	96
	6.3	Gaining Left-orders out of Ferrers-graphs	102
		6.3.1	Turning Components of Γ	102
		6.3.2	The Number of Plane Diagrams of a Lattice	105
	6.4	Determining All Non-similar Plane Diagrams of a Lattice	111

7 Conclusion and Further Work 114
7.1 Conclusion 114
7.2 Further Work 116
7.2.1 Algebraic Consequences 116
7.2.2 Minimizing the Crossing Number 116
7.2.3 Drawing Nice Diagrams 118

Chapter 1
Introduction

In the last thirty years, the development of lattice theory was boosted by the newly emerging *Formal Concept Analysis (FCA)*. Caused by its application-oriented objective, several issues gained a greater importance. In particular, the interest in nicely readable automatically layouted diagrams grew. This work is a contribution to this concern. We particularly investigate *planar* lattices, namely their characterization and representation by diagrams. Thereby, our considerations are driven by the existence of an additional order sorting the elements of a planar lattice (or poset) from left to right in contrast to the generic one which leads top-down. All structures considered will be finite.

We will describe the role of FCA as a model for handling information in Chapter 2. Lattice diagrams can be used for visualizing knowledge. We will shortly explain what a "nice" diagram may constitute and why planarity plays such an important role. Chapter 3 will lay the mathematical foundation of our work. We will recall the basic concepts of order theory, the notion of a diagram and important results about planar lattices. Furthermore we introduce FCA and *Ferrers-graphs* of *contexts* (a basic FCA structure determining a lattice).

Afterwards, we will introduce *left-relations on lattices* in Chapter 4. We will analyze its connections with *conjugate orders*, standard tools for describing planarity. Furthermore, we will give possibilities to handle left-relations more efficiently and to characterize planar contexts. In the following Chapter 5 we investigate how to actually draw a planar lattice without edge crossings. Quite helpful is the fact that left-relations on lattices can be topologically found in the shape of their respective diagrams. In the penultimate Chapter 6 we characterize planar lattices by a property of the Ferrers-graph of its context. We will give an algorithm that finds all "topologically different" plane diagrams.

Finally, we will summarize the achieved results in Chapter 7. We will give further aims seemingly interesting within the influence of this work and some hints how to approach them.

Chapter 2

Motivation

A lattice (see Definition 3.7) is *planar* (see Definition 3.16) if it can be drawn in the plane without edge crossings. In this work we want to investigate this property in detail with the help of so called *left-relations*.

However, in this chapter we want to illustrate why we are concerned with planar lattices at all. Of course, that issue is of interest just from an algebraic point of view; planarity itself and its connection to the *dimension* (see Definitions 3.5, 3.6) of a lattice are fruitful research areas (see Theorem 3.20).

In fact, our work was motivated rather from an application-oriented view. By the use of *Formal Concept Analysis* (see Section 3.5) we realize that lattice diagrams can impart information. In order to be accepted by a human user, these diagrams must look "nice" and easily readable. Although it is a hard task to mathematize human esthetic sensations, we may assume that in most cases reducing the number of edge crossings improves the diagram's quality. In particular, a lattice should be laid out without edge crossings at all, if possible.

We want to explain and reason these statements more precisely in this chapter. In Section 2.1 we will highlight the importance of diagrams for handling pieces of information. In particular, we will give a flavor of the field of *Formal Concept Analysis* including its philosophical foundation and possible applications for structuring data.

In Section 2.2 we will investigate the problems that occur by automatically drawing diagrams of discrete structures. Thereby, we refer to graphs rather than lattices since the quality of their drawings has been analyzed more specifically. We will explain the idea of *esthetic criteria* as a possibility to mathematize human esthetic conceptions and the desire of easy readability.

Finally we will illustrate in Section 2.3 why we concentrated on the esthetic criterion of *minimizing the number of edge crossings* and, in particular, on the recognition and graphical design of planar lattices as those structures that allow diagrams without edge crossings at all.

2.1 Why Do We Need Nice Diagrams?

The rapid development and increased use of computers and the worldwide spread of the internet facilitates the sharing of enormous amounts of information. However, data can quickly become too complex to organize (retrieve, store, classify, manipulate, ...) it efficiently. A new research area, namely *information science* is concerned with that issue.

Data is often presented in tables. Unfortunately, this is quite an unintuitive way of making the inherent information accessible to a human user. It seems to be useful to process the raw data into a form that fits us better. An appropriate method should apply our understanding of the principles of human thinking. A simple and well-accepted philosophical model is that of thinking in *concepts*. These are considered to be the smallest units of human thoughts. They are related by *judgements* and *conclusions*. The field of *formal concept analysis (FCA)*, introduced by R. Wille and B. Ganter in 1982, is founded on the mathematization of the interrelation between data tables (called *formal contexts*) and respective concepts (called *formal concepts*).

In that model, our knowledge (or parts of it) is based on a set of objects, a set of attributes and an incidence relation stating which objects possess which attributes. This structure is called a formal context. A formal concept consists of an *extension*, i.e. a set of objects and an *intension*, i.e. a set of attributes. Thereby, all objects possess all attributes (and do not have further attributes in common) and, vice versa, all attributes apply to all objects (and no other object possesses all of them). Concepts can be ordered into the *concept hierarchy*, on the one end one finds the more special concepts consisting of few objects and many common attributes, on the other the more general concepts occur, consisting of many objects sharing only few attributes.

Of course, this is a simplification of the rather complex human thinking processes. The addition of the word "formal" refers to that fact: a mathematical field can not at all be a model for such a complex interaction. However, this theory affords manifold possibilities to process and represent information in a meaningful and helpful manner. Mathematically, formal concept analysis is a section of applied lattice theory. The concept hierarchies turn out to be complete lattices (see Definition 3.7) that are called *concept lattices*.

By drawing diagrams of concept lattices, we enable the user to discover visually the information that is hidden in the appropriate formal context. This aspect seems to be another advantage of the use of diagrams in general: seeing information (if one realizes the structure of the diagram logically) is always better than reading it. Additionally, it is possible to explore the "neighborhood" of an interesting concept, a feature that is hardly provided by usual data base platforms.

Chapter 2: Motivation

There already exist several applications using FCA for representing knowledge: The programs MAILSLEUTH [EDB04] and its free clone HIERMAIL, that are both derived from CEM [CS00], provide email clients that allow to organize ones emails conceptually. IMAGESLEUTH [DVE06] and CAMELIS [Fer07] are attempts to organize image data bases with the help of FCA. SEARCHSLEUTH [DE07] and FOOCA [Kös06] are plug-ins for internet search engines and organize the results of a respective query in the same manner. More generally, D-SIFT [DWE05] and TOSCANAJ [BHS02] allow to process arbitrary data bases. There exist FCA-based tools for managing data bases of many more areas like scientific articles (BIBSONOMY [JHSS07]) or media in a library. Eventually, even the complete file system of a computer could be organized this way; instead of putting documents into folders, one could apply attributes to them and then search via these attributes.

We want to pick a very specific application named SURF MACHINE [DE05] that indicates suitable surfing spots in western Australia due to dominant wind and wave directions. See Figure 2.1 for an illustration.

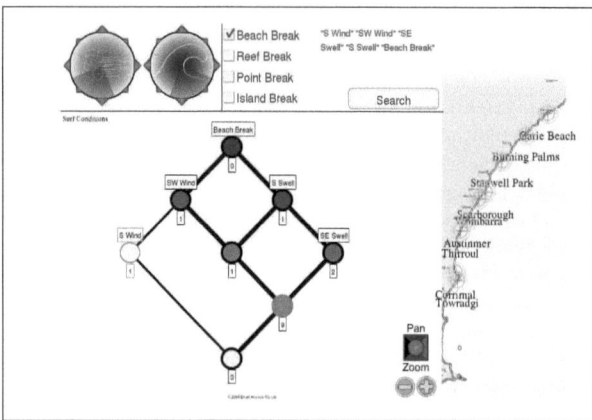

Figure 2.1: SURF MACHINE - an example of an FCA based and diagram driven information system.

After entering the probable weather conditions according to the forecast in the upper part of the window, the program provides a diagram with surfing spots that suit better the lower one descends. Ideally, the elements, i.e. surf spots, contained in the bottom node supply good conditions for all chosen inputs. If such an element does not exist (which is indicated by the 0 below that node) one can decide to drop one (or several) of the desired properties by running up one or several diagram edges. In our particular case the decision was made to ignore the attribute "southerly winds" by moving to the upper right. By clicking on the respective node, the 9 included surf spots are shown on the right.

Although SURF MACHINE seems to be a toy example only, it points out some of the many features offered by the FCA framework. It helps, for example, to *search* for good results, *decide* what is the best result, *investigate* what conditions lack good surfing spots or *identify* which spots have the same characteristics.

2.2 What Is a Nice Diagram?

Recently, newly developed FCA-based information systems (e.g., SEARCH-SLEUTH, CAMELIS) tend not to use diagrams for presenting data anymore. In our opinion, this is due to two facts. Firstly, diagrams are becoming too big to be understood by a human user. Secondly, it is quite complicated to create an algorithm designing their layout automatically. Although there already exist some appropriate applications (for instance the widely used CONEXP [Yev], the diagram browser CERNATO used in TOSCANAJ, [BHS02] or the online application JALABA [Fre]), none of them is completely satisfactory. The first problem can be solved to a certain degree by employing *nested line diagrams* [GW99]. However, our concern is to tackle the second problem, i.e. to find strategies to draw diagrams automatically with better quality.

Obviously, this leads to the question what distinguishes a nice diagram? Since it is used by a human user, it should be accepted by him. This means that, it should *correspond to his esthetic sensations* and furthermore be *easily readable*. Apparently, we need criteria that can be measured in a diagram to decide how much these two desires are satisfied.

In the *graph drawing* community (see e.g., [DETT99]) so called *esthetic criteria* were introduced for that purpose. They can be understood as optimization tasks whose fulfillment increases the diagrams quality. Examples include

1. *minimize the number of edge crossings,*
2. *maximize the least angle between incident edges,*
3. *minimize the number of different slopes,*
4. *display symmetries,*
5. *place the nodes onto an orthogonal grid with minimum size (w.r.t. the edge length of the grid).*

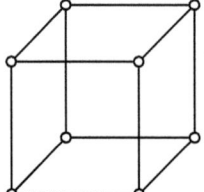 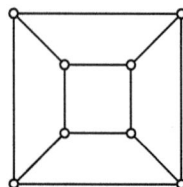 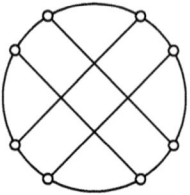

Figure 2.2: Three diagrams of the cube.

Moreover, there exist so called *drawing conventions* [CT94] describing the general shape of a diagram. For us, the most important examples include *layered drawings* and *additive drawings* (see Definitions 3.12 and 3.13). They can also be considered as esthetic criteria or; at least their usage emphasizes several criteria over others.

While laying out a diagram, one often has to prefer some criteria over others depending on the desired properties of the result.

In Figure 2.2, three visualizations of a cube, seen as a simple graph, are depicted. The left diagram complies with our imagination of a cube being projected into the plane. The one in the middle gets by without edge crossings, while the one on the right emphasizes inherent symmetries, namely the node transitivity[1].

In addition to the esthetic criteria introduced for general graphs, further conditions were developed in the FCA community to improve the quality of lattice diagrams in particular. Examples include the principles of a *functional layout* [CDE06], the *maximization of the conflict distance* [Gan04] or the convention of *additive diagrams* [Sko92, GW99]. Of course, design aspects beyond mathematics must be regarded as well, the *look and feel* plays an important role in acceptance of lattice diagrams [EDB04].

2.3 Why Planarity?

After collecting all these constraints one encounters another question: *Which esthetic criteria are important?* Until now we just introduced them without giving evidence for their necessity. Of course, such a verification can be done only empirically, since it is obviously not possible to mathematically measure the quality of a diagram in terms of utility for human users.

As far as we know, only two empirical studies [Pur97, PAC00] were undertaken for graph diagrams (and none for lattices). In the first, the test persons

---

[1] That is, for every pair of nodes v, w, there exists an automorphism φ mapping v onto w.

2.3 Why Planarity?

had to perform tasks on the given diagrams, while in the second they had to choose which diagrams they preferred. Of the tested criteria, *minimizing the number of edge crossings* turned out to play the most important role (the other included criteria were *minimize number of bends, place nodes on orthogonal grid, maximize least angle between incident edges, display symmetries*).

When finding strategies to minimize the number of edge crossings, it is an obvious prerequisite to be able to distinguish between planar and non-planar lattices and to draw the latter without edge crossings at all. This was ultimately the driving force for working on this topic at all. It turned out to be so fruitful that we concentrated nearly exclusively on planarity and left the problem of analytically minimizing the number of edge crossings and more generally the automatic layout of lattices for later research.

Chapter 3
Preliminaries

In this chapter we will recall the theory and refer to some results that form the background of our work.

At the very beginning in Section 3.1 we will give an elementary introduction into some aspects of order theory. In particular, we will define lattices as ordered sets with additional properties. Moreover, some lesser known but nevertheless basic notations will be recalled.

Section 3.2 will address *diagrams of lattices*. We will define this concept first, since a careful distinction between lattices and diagrams (although both notions are used synonymously in many publications) is necessary in Chapter 5. Afterwards we will describe some special classes of diagrams that are common for drawing lattices.

Section 3.3 will subsume some important results aiming at the characterization of *planar lattices*. In particular we will highlight the concepts of *conjugate orders* as the *left relations on lattices* in Chapter 4 are closely related to it.

In Section 3.4 we will note some remarks about the *geometry of planar lattices* [KR75]. These will be used in particular in Chapter 5 when we actually draw planar lattices.

As mentioned in Chapter 2 this work is motivated by the applications arising in *formal concept analysis*. Therefore we will recapitulate some basic assertions of this field, in particular its connection to lattice theory, in Section 3.5. Additionally, the interrelation of contexts and respective concept lattices will be needed in Section 4.4. There we will describe the planarity of a lattice by its *standard context*.

Finally, Section 3.6 will be concerned with *Ferrers relations*. We will introduce the *Ferrers graphs* which will be the key point for the characterization of the set of all plane diagrams in Chapter 6.

We omitted to recall some basics about *graph theory* that we use in Chapter 6. We refer to a standard monograph (e.g., [Die96]) for the interested reader.

3.1 Basic Definitions

Order theory, and its branch lattice theory, are based on the definition of an *ordered set*.

Definition 3.1 *[Bir67] Let P be a set and $R \subseteq P \times P$ a binary relation[1]. Let Δ_P denote the identity relation on P. If R satisfies the three conditions*

1. *R is* reflexive, *i.e.* $\Delta_P \subseteq R$,

2. *R is* antisymmetric, *i.e.* $R \cap R^{-1} \subseteq \Delta_P$,

3. *R is* transitive, *i.e.* $R \circ R \subseteq R$,

then we call R order relation *or shortly* order *on P and the pair $\underline{P} := (P, R)$* ordered set *or shortly* poset.
If the relation R fulfills condition 3 and, instead of 1 and 2, the condition

1'. R is asymmetric, *i.e.* $R \cap R^{-1} = \emptyset$.

then R is called strict order *on P.* ◊

An important subclass of ordered sets are *linear orders*. Usually, order relations are denoted with the symbol $\leq$, which is well known for denoting the standard linear order on numbers.

Definition 3.2 *[Bir67] Let $\underline{P} = (P, R)$ be an ordered set. If the relation R satisfies the condition*

4. *R is* connex, *i.e.* $R \cup R^{-1} = P \times P$,

then R is called linear order *and the pair (P, R) is called* chain. *If R is a strict order fulfilling $R \cup R^{-1} = (P \times P) \setminus \Delta_P$ then R is called* strict linear order. ◊

Two elements $p, q \in P$ which are in relation R, i.e. satisfy either $p\,R\,q$ or $q\,R\,p$, are called *comparable*. Otherwise they are *incomparable*. The pairs of incomparable elements will play a fundamental role in the course of this work, therefore we define formally:

Definition 3.3 *[Grä98] Let $\underline{P} = (P, R)$ be an ordered set. The relation*

$$\| := (P \times P) \setminus (R \cup R^{-1})$$

is called incomparability relation *on $\underline{P}$.* ◊

---
[1] We assume that the reader is familiar with the notion of a *relation*.

For drawing diagrams of ordered sets, we additionally need the notation of the *neighborhood relation*. It contains, as the name suggests, only those pairs of comparable elements p and q which do not have another element "in-between" them, i.e. no element r different from p and q fulfilling $p\,R\,r\,R\,q$. In a diagram one usually encounters edges between neighbored elements only. This is due to better readability.

Definition 3.4 *Let $\underline{P} = (P, R)$ be an ordered set and $\tilde{R} := R \setminus =_P$.*

1. *[BLS99] With $(P, R \cup R^{-1})$ we denote the* comparability graph *of $\underline{P}$.*

2. *[DM41, Ore62, Pla76] The relation $\prec := \tilde{R} \setminus (\tilde{R} \circ \tilde{R})$ is called* neighborhood relation *or* cover relation *or* transitive reduction *of P. If $a \prec b$ holds then we call b* upper cover *of a and a* lower cover *of b. The pair $(P, \prec)$ is called* graph[2] *of $\underline{P}$.* ◇

The purpose of introducing graphs of ordered sets is to draw better diagrams. An example is depicted in Figure 3.1. Obviously, the left diagram is harder to read. Although we omitted all loops, it still contains redundant edges. A human user will instinctively complete the diagram on the right to its transitive closure, i.e. the relation R itself.

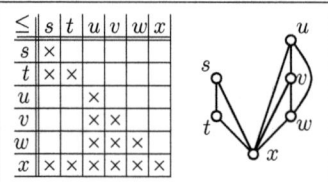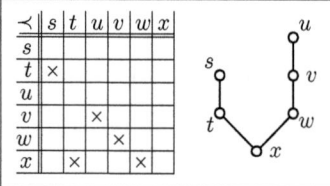

Figure 3.1: Comparison of an ordered set (left) and its graph (right). In the left diagram all loops, i.e. all edges of the form (p, p), are removed already.

In the next section we will see that the concept of planarity is fundamentally connected to that of the *order dimension*. In particular, in Section 5.3 we will use *realizers* of a planar lattice to construct a plane diagram of it.

Definition 3.5 *[DM41] Let $\underline{P} = (P, R)$ be an ordered set. Let I be an index set and $\{R_i \mid i \in I\}$ be a family of linear orders on P whose intersection equals R, i.e. $R = \bigcap_{i \in I} R_i$. Then $\{R_i\}_{i \in I}$ is called* realizer[3] *of $\underline{P}$.*

The order dimension *(or shortly* dimension*) of $\underline{P}$ is the smallest cardinality $|I|$, s.t. $\{R_i \mid i \in I\}$ is a realizer of $\underline{P}$.* ◇

---
[2] Ore used the notion *basis graph* and Platt *covering graph* instead. By a *graph* of an ordered set $\underline{P}$ Ore denoted any directed acyclic graph whose transitive closure yields $\underline{P}$.

[3] The existence of a realizer for arbitrary posets was evidenced in [DM41].

3.1 Basic Definitions

See Figure 3.2 for an example of an ordered set together with a realizer. We have seen that the order dimension is defined via the intersection of linear orders. There exists another concept of dimension in order theory called *product dimension*. Its underlying idea is an embedding into a product of chains. For a visual comparison between both constructions we refer to Figure 3.2.

Definition 3.6 *[Ore62] Let $\underline{P} = (P, \leq)$ be an ordered set. The product dimension $pdim(\underline{P})$ of $\underline{P}$ is the smallest cardinality n, s.t. there exists an order embedding[4] of $\underline{P}$ in a product[5] of n chains.* ◇

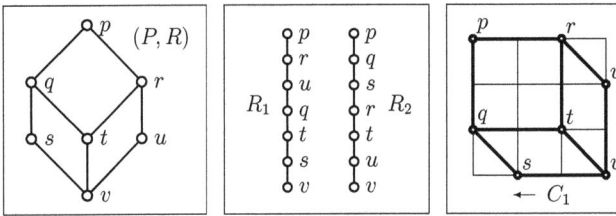

Figure 3.2: An ordered set (P, R) (left) together with a realizer $\{R_1, R_2\}$ (middle) and an embedding into the product of two chains C_1 and C_2 of length 4. The arrows indicate the direction of the greater elements w.r.t each chain. Order dimension and product dimension both equal 2. All structures are given by means of their diagrams.

An *upper bound* of a subset $X \subseteq P$ of an ordered set $\underline{P} = (P, \leq)$ is an element y satisfying $x \leq y$ for all elements $x \in X$. Dually a lower bound is an element z with $z \leq x$ for all $x \in X$. We call $y = \bigvee X$ *least upper bound* if it is lesser or equal to every upper bound y' of X. Dually $z = \bigwedge X$ is a *greatest lower bound* of X if it is greater or equal to every lower bound z' of X. The order theoretic definition of a lattice is based on the existence of those bounds.

Definition 3.7 *[Bir67] Let $\underline{\mathfrak{V}} = (\mathfrak{V}, \leq)$ be an ordered set. If, for every two elements v and w, the least upper bound $x \vee w$ (called join) and the greatest lower bound $x \wedge w$ (called meet) exist then $\underline{\mathfrak{V}}$ is called lattice. If the least upper bound and greatest lower bound exist for every subset X of $\underline{\mathfrak{V}}$ then $\underline{\mathfrak{V}}$ is called complete lattice.* ◇

---

[4] An *order embedding* is a mapping $\varphi : P \to Q$ between two ordered sets (P, R_1) and (Q, R_2), s.t. $p_1 R_1 p_2 \iff \varphi(p_1) R_2 \varphi(p_2)$ holds for all $p_1, p_2 \in P$

[5] We recall the finite case only: The *product* of a set $\{(P_i, R_i) \mid i \in \{1, \ldots, n\}\}$ of ordered sets is the ordered set $(P_1 \times \ldots \times P_n, R)$ with $(p_1, \ldots, p_n) R(q_1, \ldots, q_n) : \iff p_i R_i q_i$ for all $i \in \{1, \ldots, n\}$.

The following facts are well-known and easy to prove:

1. Every finite lattice is a complete lattice.

2. Every chain is a lattice.

3. Every complete lattice contains a top element 1 and a bottom element 0.

Next we want to highlight some special elements in lattices. They gain importance in particular in Formal Concept Analysis.

Definition 3.8 *[Grä98] Let $\mathfrak{V} = (\mathfrak{V}, \leq)$ be a finite lattice.*

1. An element $m \in \mathfrak{V}$ fulfilling

$$m = v \wedge w \implies m = v \text{ or } m = w$$

for all elements $v, w \in \mathfrak{V}$ is called $\bigwedge$-irreducible[6].

2. An element $j \in \mathfrak{V}$ fulfilling

$$j = v \vee w \implies j = v \text{ or } j = w$$

for all elements $v, w \in \mathfrak{V}$ is called $\bigvee$-irreducible.

3. An element is called doubly-irreducible *if it is both $\bigvee$ and $\bigwedge$-irreducible.*

The set of $\bigwedge$-irreducibles is denoted by $M(\mathfrak{V})$ and the set of $\bigvee$-irreducibles by $J(\mathfrak{V})$. ◇

Every element of a finite lattice can be represented as an infimum of $\bigwedge$-irreducibles. A set fulfilling this property is called *infimum-dense*. The dual holds for the join operation. The sets of $\bigwedge$-irreducibles and of $\bigvee$-irreducibles can be recognized easily in a diagram, as the following lemma states.

Remark 3.9 *[GW99] Let $\mathfrak{V}$ be a finite lattice. An element $m \in \mathfrak{V}$ is a $\bigwedge$-irreducible if and only if it has exactly one upper cover. Dually, an element $j \in \mathfrak{V}$ is a $\bigvee$-irreducible if and only if it has exactly one lower cover.*

We will denote by v^* and v_* the set of upper and lower covers of an element v, with m^* the unique upper cover of a $\bigwedge$-irreducible m and with j_* the unique lower cover of an $\bigvee$-irreducible j. The set $M(v) := \{m \in M(\mathfrak{V}) \mid v \leq m\}$ contains all $\bigwedge$-irreducibles greater or equal to a lattice element v.

---

[6]We chose here this denotation in contrast to the more common ∧-*irreducible*.

3.2 Diagrams of Ordered Structures

In many cases it is not necessary to distinguish between a lattice, or an ordered set, and a diagram representing it. Therefore, it is common to consider both concepts synonymously. However, for our purpose we explicitly need an unambiguous distinction between them since we actually want to draw diagrams of given lattices. In our everyday experience, a diagram of an ordered structure consists of little circles (sometimes also rectangles or other polygons) and arcs connecting adjacent elements. It is common to represent the order by an increasing y- or x-coordinate in order to avoid the arrow symbols on edges.

More formally, the diagram of an ordered set $\underline{P}$ is a *graph diagram*[7] of the graph of $\underline{P}$ meeting the upward-drawing-constraint:

Definition 3.10 *[Qua73, KR75] Let $\underline{P} = (P, \leq)$ be an ordered set with neighborhood relation $\prec$. A diagram (or representation [KR75]) $\mathrm{pos}(\underline{P})$ of $\underline{P}$ is a mapping*[8]

$$\mathrm{pos}: P \cup \prec \; \to \; \mathbb{R}^2 \cup \mathcal{P}(\mathbb{R}^2)$$

meeting the following conditions for all $v, w, z \in P$:

1. *Elements $v \in P$ are mapped injectively to points in the Euclidean plane:*

   $$\mathrm{pos}\,|_P:\; v \mapsto \mathrm{pos}(v) = (x(v), y(v)) \in \mathbb{R}^2 \text{ is an injection.}$$

2. *The upward drawing constraint or Hasse-constraint is satisfied:*

   $$v < w \implies y(v) < y(w).$$

3. *Adjacent nodes $v \prec w$ are connected by upward arcs:*

   $$\mathrm{pos}\,|_\prec:\; (v, w) \mapsto \{(x_{vw}(y), y) \mid y \in [y(v), y(w)]\} \subseteq \mathbb{R}^2,$$

   where $x_{vw} : [y(v), y(w)] \to \mathbb{R}$ is a continuous function satisfying both $x_{vw}(y(v)) = x(v)$ and $x_{vw}(y(w)) = x(w)$.

4. *The diagram is conflict free, i.e. no node is situated on a non-incident line:*

   $$\mathrm{pos}(v) \in \mathrm{pos}((w, z)) \implies v = w \text{ or } v = z.$$

The elements of $\mathrm{pos}(P) := \{\mathrm{pos}(v) \mid v \in P\}$ are called (diagram) points or nodes, the elements of $\mathrm{pos}(\prec) := \{\mathrm{pos}(v, w) \mid v, w \in P, v \prec w\}$ are called diagram edges. ◊

---

[7]This is, a diagram in the sense of graph theory.
[8]With $\mathcal{P}(\mathbb{R}^2)$ we denote the set of subsets of $\mathbb{R}^2$.

Commonly, *line diagrams* are employed, if possible. Then the edges are just straight lines. There evolved many names for this structure, we will stick to the term used in [GW99] since in our opinion it is the most intuitive synonym.

Definition 3.11 *[KR75] A line diagram (also called* embedding [KR75], Hasse diagram [BFR71] *or simply* diagram [Bir67]*) of an ordered set $\underline{P}$ is a diagram (as previously defined), where*

$$\mathrm{pos}((v,w)) = \{t \cdot \mathrm{pos}(v) + (1-t) \cdot \mathrm{pos}(w) \mid t \in [0,1]\}.$$

holds for all elements $v \prec w$. This is, all diagram edges are straight line segments. ◊

There exist some more specific conventions to draw diagrams, as mentioned already in Section 2.2. Here we want to introduce the most important ones.

Layer diagrams are characterized by a *layer assignment function* which arranges the nodes of lattice elements onto horizontal lines called layers.

Definition 3.12 *[STT81, DETT99] Let $\underline{P} = (P, \leq)$ be an ordered set. A map $p: P \to \mathbb{R}$ is called* layer assignment function *if $v < w \implies p(v) < p(w)$ holds for all pairs of elements $v, w \in P$. A diagram $\mathrm{pos}(\underline{P})$ is called* layer diagram *w.r.t. p if $\mathrm{pos}: v \mapsto (x(v), p(v))$ holds for all $v \in P$, i.e. if the y-coordinate of every element equals its layer.* ◊

Of course, every diagram possesses a layer assignment function p determined by $p(v) = y(v)$ for all elements of $\underline{P}$. However, the purpose of that drawing convention is to place many nodes onto each layer followed by a crossing number minimization by ordering the nodes in each layer [DETT99]. A common assignment function is the *longest path layering* [ES90], where each element of $\underline{P}$ is mapped onto the length of the longest path to any maximal element of $\underline{P}$. Lattice drawing algorithms applying the layer assignment convention include JALABA [Fre] and CONEXP [Yev].

Additive diagrams evolved in the FCA-community in order to draw diagrams of concept lattices. In particular distributive (or "nearly distributive") lattices look like drawn on an n-dimensional grid [Sko92].

Definition 3.13 *[GW99] Let $\underline{P} = (P, \leq)$ be an ordered set. A set representation* rep *is an order embedding of $\underline{P}$ into the powerset $\mathcal{P}(X)$ of a set X.*

An additive diagram *is a line diagram of $\underline{P}$ determined by a set representation* rep *and a* grid projection vec $: X \mapsto \mathbb{R}^2$ *assigning a vector with positive y-component to each element of X, s.t. the equation*

$$\mathrm{pos}(p) = n + \sum_{x \in \mathrm{rep}(p)} \mathrm{vec}(x)$$

3.2 Diagrams of Ordered Structures

holds for all elements p of P. Thereby, n is an arbitrary vector allowing to shift the whole diagram in the plane.   ◊

Again, this definition seems somehow meaningless since every line diagram of a lattice $\underline{\mathfrak{V}} = (\mathfrak{V}, \leq)$ can be interpreted as an additive diagram. Just choose $X = \mathfrak{V}$ and

$$\text{rep}(v) = \{w \in \mathfrak{V} \mid w \geq v\} \quad \text{and} \quad \text{vec}(v) = \text{pos}(v) - \sum_{x \in \text{rep}(v) \setminus \{v\}} \text{vec}(x).$$

The aim of employing additive diagrams is to better display symmetries of the lattice. This is done by choosing a smaller set X. Usually one defines $X = M(\underline{\mathfrak{V}})$ or $X = J(\underline{\mathfrak{V}})$, which leads to *attribute* or *object additivity*. In particular distributive lattices (or "nearly distributive" ones) are drawn quite nicely then. Therefore this technique is used widely in the FCA-community in algorithms like TOSCANAJ [BHS02] or CONEXP [Yev][9].

In the following we will formally define the attribute additive drawing convention. It does not actually result in an additive diagram in the sense of Definition 3.13 since the mapping rep : $v \mapsto M(v)$ is not an order embedding, but an anti-order embedding. However, if we assign vectors with negative second component to the elements of $M(\underline{\mathfrak{V}})$ we still gain a diagram satisfying the Hasse constraint.

Definition 3.14 *[GW99, CDE06] Let $\underline{\mathfrak{V}} = (\mathfrak{V}, \leq)$ be a lattice. A line diagram pos($\underline{\mathfrak{V}}$) is called* attribute additive *if there is a map* vec : $M \mapsto \mathbb{R}^2$, *such that the equation*

$$\text{pos}(v) = \sum_{m \in M(v)} \text{vec}(m)$$

holds for all elements $v \in \mathfrak{V}$.   ◊

We have to mention that the term *attribute* evolved, together with the whole definition, in FCA. There, the role of the $\bigwedge$-irreducibles is taken by the *irreducible attributes* of a context. So the last definition is a restriction to the original one in [GW99] since all reducible attributes are mapped by vec to the zero vector. However, we think that our (also provided in [CDE06]) sloppy interpretation is very useful since it aims at the layout of *grid structures* [Sko92]. These obviously help to increase the readability of the underlying lattice and are therefore desired by us. An example for such a drawing can be found in Figure 3.3.

Of course, it is feasible to develop diagram types mixing the constraints of the layer and the additive convention. The *hybrid diagrams* introduced by

---
[9]In fact, CONEXP provides several drawing conventions.

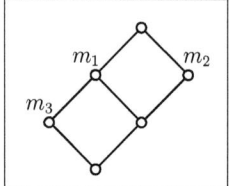 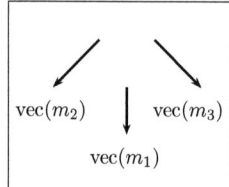 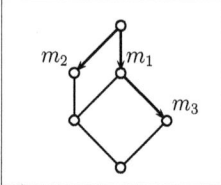

Figure 3.3: A lattice (left, given by a diagram of it) together with a grid projection vec in the middle. The emerging diagram is depicted on the right.

R. Cole [Col00, CDE06] are one example. Only the x-coordinates of lattice elements are calculated via the attribute additivity, the y-coordinates however are determined by layer assignment (in fact the papers present only longest path layering). SURF MACHINE [DE05] uses this drawing convention.

Another attempt is oriented rather towards the aim of layering attribute additive diagrams. An easy solution is to supply a fixed y-coordinate to the grid projection function of all $\bigwedge$-irreducible elements. This convention is quite common for manual drawings of many lattices. In this case the layer assignment function p is, for a fixed number $c \in \mathbb{R}$, given by

$$p(v) = c \cdot |M(v)|.$$

Definition 3.15 *Let* $\operatorname{pos}(\mathfrak{V})$ *be an attribute additive diagram of the lattice* $\mathfrak{V}$. *If the grid projection function* vec *inducing* $\operatorname{pos}(\mathfrak{V})$ *satisfies the condition*

$$\exists c \in \mathbb{R} \ \forall m \in M : \operatorname{vec}(m) = (x(m), c)$$

then $\operatorname{pos}(\mathfrak{V})$ *is a* layered attribute additive diagram *of* $\mathfrak{V}$. ◊

The diagram depicted on the right picture of Figure 3.3 is even layered attribute additive. The top element is situated on the top layer. The elements m_1 and m_2 (fulfilling $|M(m_1)| = |M(m_2)| = 1$) are on the second layer, the elements m_3 and $m_1 \wedge m_2$ (fulfilling $|M(m_3)| = |M(m_1 \wedge m_2)| = 2$) are on the third and the bottom element on the fourth.

Contrary to the well-known previous Definitions 3.12 and 3.13, the layered attribute additivity is not theoretically observed yet although practically used widely. Our interest in this work is to draw plane diagrams of lattices. Hence, we will come back to this convention in Subsection 5.3.

3.3 Planar Lattices

In this section we recall some important characterizations of planar lattices. As already mentioned, our interest is focused particularly on the connection with *conjugate orders* and the *order dimension* since this is of fundamental importance to our further work.

At first we recapitulate the well-known definition of planarity. In contrast to the same notion on graphs, we demand all diagram edges to be directed upward (see condition 2 of Definition 3.10).

Definition 3.16 *[Bir67]* *A lattice $\mathfrak{V}$ is upward planar or shortly planar if it possesses a plane diagram, i.e. a diagram $\mathrm{pos}(\mathfrak{V})$ without an edge crossing.* ◊

The notion of an *edge crossing* is quite intuitive. We will further investigate this concept in section 3.4. Every *planar graph* possesses a plane diagram without edge bends due to the Theorem of Fary [Far48]. An equivalent statement for posets has been achieved by Kelly [Kel73]: Every (upward) planar ordered set has a plane (line) diagram. This result was even extended in [Kel87] in the following way: A poset $\underline{P}$ is planar if and only if there exists a diagram[10] $\mathrm{pos}(\underline{P})$ without edge crossings, where the diagram lines $\mathrm{pos}(a,b)$ between two elements $a \prec b$ are arbitrary arcs that are stretched between the y-coordinates of a and b. We will further investigate this issue in Chapter 5 and answer the questions whether an arbitrary planar lattice possesses a plane *attribute additive* and a plane *layer* diagram respectively.

A *conjugate order* on an ordered set $(P, \leq)$ is an additional order acting "from left to right" (using the common principle to sort comparable elements "bottom-up").

Definition 3.17 *[DM41]* Let $\underline{P} = (P, \leq)$ be an ordered set and $L_c \subseteq P \times P$.

1. L_c *is a* conjugate relation *if $L_c \cup L_c^{-1} = \|$.*

2. L_c *is a* conjugate order *if additionally L_c is a strict order.* ◊

A conjugate order is an additional order on a lattice. In Chapter 4 we will construct and describe them with the help of *left-relations*.

An interesting property of an ordered set $\underline{P} = (P, \leq)$ possessing a conjugate order is the following: Every two elements $p, q \in P$ are in exactly one of the five relations $<, >, L_c, L_c^{-1}$ and $=$. Hence, we may write

$$<_{\underline{P}} \dot{\cup} >_{\underline{P}} \dot{\cup} L_c \dot{\cup} L_c^{-1} \dot{\cup} =_{\underline{P}} \ = \ P \times P.$$

---
[10] Although this is not a diagram in the sense of Definition 3.10, we keep that name for convenience

Conjugate orders define linear extensions of an ordered set and thus give a link to its dimension. Vice versa, realizers also determine conjugate orders:

Lemma 3.18 *Let $\underline{P} = (P, \leq)$ be a partial order.*

[DM41] Every realizer $\{P_1, P_2\}$ of $\underline{P}$ determines two conjugate orders $P_1 \backslash \leq$ and $P_2 \backslash \leq$ satisfying $(P_1 \backslash \leq)^{-1} = P_2 \backslash \leq$.

[Gol80] For every conjugate order L_1, the relation $L_2 = L_1^{-1}$ is a conjugate order, too. The pair $\{L_1 \cup \leq, L_2 \cup \leq\}$ is a realizer of $\mathfrak{V}$.

Finally, as a preparation of Theorem 3.20, we introduce *interval-inclusion-lattices*. See Figure 3.4 for an example.

Definition 3.19 *[GY99] A lattice $\mathfrak{V} = (I, \subseteq)$ is an* interval-inclusion-lattice (IIL) *if $I \subseteq \mathcal{I}(\mathbb{R})$ is a set of intervals over $\mathbb{R}$ that forms together with the set inclusion relation $\subseteq$ a lattice.* ◇

The following theorem is a collection of results provided by several authors. Dushnik and Miller [DM41] first showed the equivalence of the conditions 2, 3 and 5. Indeed they proved that more generally for arbitrary ordered sets. The equivalence of 1 and 2 was stated in [Bir67] (p.32, ex. 7c). The "$\Rightarrow$" part was proved by Kelly and Rival in [KR75] (see Proposition 3.32) the construction for showing the "$\Leftarrow$" part can be found for instance in [KT82]. However, we will recall it in Theorem 5.6. The equality of 3 and 4 was (more generally for ordered sets) given by Ore in [Ore62]. Finally, the equality between 2 and 6 is immediate and was first mentioned in [GH62]. We should remark that the well known theorem of Baker, Fishburn and Roberts [BFR71], stating the equivalence of 1 and 3 is a subsumption of the above-mentioned results.

Theorem 3.20 *[BFR71, Bir67, DM41, KR75, Ore62, GH62] Let $\mathfrak{V} = (\mathfrak{V}, \leq)$ be a finite lattice. The following conditions are equivalent:*

1. *$\mathfrak{V}$ is planar.*

2. *There exists a conjugate order L_c on $\mathfrak{V}$.*

3. *The order dimension $dim(\mathfrak{V})$ of $\mathfrak{V}$ is at most two*[11].

4. *The product dimension $pdim(\mathfrak{V})$ of $\mathfrak{V}$ is at most two.*

5. *There exists a representation of $\mathfrak{V}$ by an IIL.*

6. *The graph $(P, \|)$ is a comparability graph.*

---

[11] Contrary, a planar poset may have an arbitrary finite order dimension, see [Kel81]).

3.3 Planar Lattices

Sketch of the proof: We want to give the main ideas leading from the existence of a conjugate order L_c on $\mathfrak{V}$ to the validity of the other conditions.

Together with L_c, also L_c^{-1} is a conjugate order. Both $L_c \cup \leq$ and $L_c^{-1} \cup \leq$ are linear extensions of $\leq$ (see Lemma 3.18) whose intersection equals $\leq$. Therefore the order dimension is at most two. We can provide two order embeddings l and r of $L_c \cup \leq$ and $L_c^{-1} \cup \leq$ into the linear ordered sets $\underline{R_1} := (\mathbb{R}^+, <)$ and $\underline{R_2} := (\mathbb{R}^+, <)$ respectively. A line diagram defined by

$$\text{pos} : \mathfrak{V} \mapsto \mathbb{R}^2 \qquad \text{with pos}(v) := (l(v), r(v))$$

is plane (see the proof of Theorem 5.6 for details). Additionally it provides an embedding into the product of the two chains $\underline{R_1}$ and $\underline{R_2}$. If we on the other hand define a mapping

$$\varphi : \mathfrak{V} \mapsto \mathcal{I}(\mathbb{R}) \qquad \text{with } \varphi(v) = (-r(v), l(v))$$

into the set of intervals over $\mathbb{R}$ then φ is an isomorphism into an IIL.

See Figure 3.4 for a visualization of the maps pos and φ of the lattice $\mathfrak{V}$ given on the right. We can find a conjugate order L. Both relations $L_< := L \cup <$ and $R_< := L^{-1} \cup <$ are linear orders on $\mathfrak{V}$ (depicted on the upper left). The embeddings l and r into the real numbers together with an appropriate order can be seen on the upper right. The mapping pos supplying a plane line diagram pos($\mathfrak{V}$) is depicted in the lower left.

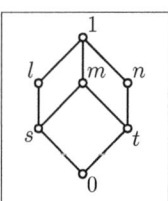

Such a diagram, where a directed path between u and v occurs if and only if $x(u) \leq x(v)$ and $y(u) \leq y(v)$ holds, is called *dominance drawing* [BCB+92, CT94]. We will further observe this kind of diagram in Section 5.3. Finally a mapping φ into a set of intervals over $\mathbb{R}$ is given on the lower right of the figure.

Besides this fruitful bundle of results several other characterizations concerning the planarity of lattices were discovered. Although we do not need all these achievements in our further work we will mention them in order to have a broader view on the topic.

In Section 3.1 we recalled the term of doubly irreducible elements. If an ordered set can be reduced to the two-elemental chain by a consecutive deletion of doubly irreducible elements, it is called *dismantlable*. See Figure 3.5 for an example. Firstly introduced by Rival in [Riv74], we refer to the following definition:

Definition 3.21 *[KR75] A finite ordered set $\underline{P} = (\{p_1, \ldots, p_n\}, \leq)$ is dismantlable, if for all $i \in \{1, \ldots, n\}$ the element p_i is doubly irreducible in the subposet $(\{p_1, \ldots, p_i\}, \leq)$.* ◇

Chapter 3: Preliminaries

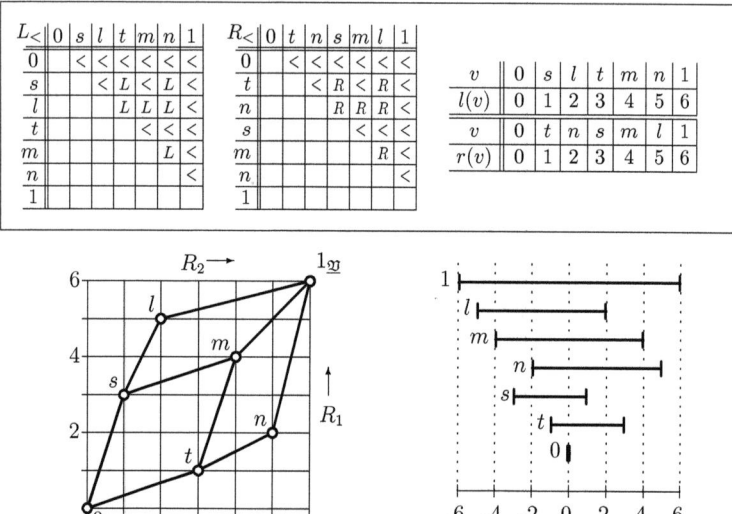

Figure 3.4: The construction of two realizers $L_<$ and $R_<$ of a lattice $\mathfrak{V}$ out of a conjugate order L supplies a plane diagram $\mathrm{pos}(\mathfrak{V})$ and an isomorphism φ into an interval inclusion lattice.

It is easy to see that a dismantlable finite bounded ordered set is a lattice [KR75]. It is surprising, however, that every plane diagram of a planar ordered set possesses a doubly-irreducible element on its *boundary*[12]. This result was first proved in [BFR71] and formulated tighter in [KR75]. It gives rise to several statements about planar lattices. The first immediate consequence is:

Theorem 3.22 *[KR75] A planar finite bounded ordered set is dismantlable.*

For special classes of lattices, a more precise characterization is possible. E.g. a distributive lattice is planar if and only if it is dismantlable, see [Qua73].

With the help of Theorem 3.22 it is possible to transform the issue of the planarity of a lattice to that of graphs[13].

---
[12]This term is to be understood in the natural (topological) way, see [KR75] for a formal definition.

[13]Planar graphs are well investigated, see [HT74] and [LEC67] for algorithms constructing plane diagrams in linear time.

3.3 Planar Lattices 25

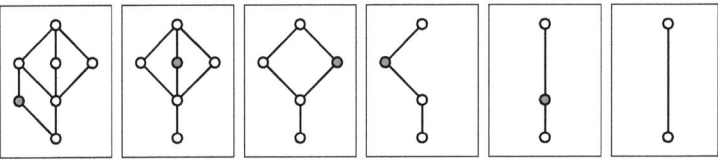

Figure 3.5: The process of dismantling a lattice. The doubly irreducible elements that are going to be deleted are shaded.

Theorem 3.23 *[Pla76] A finite lattice $\mathfrak{V} = (\mathfrak{V}, \leq)$ is planar if and only if the undirected graph derived from its graph $(\mathfrak{V}, \prec)$ by ignoring the orientation of the edges and adding an edge between bottom and top element is planar.*

The next theorem appears first as an exercise (p.32 Ex. 7a) in [Bir67]. Proofs can be found in [Qua73, Pla76].

Theorem 3.24 *[Bir67] A planar finite bounded ordered set is a lattice.*

In graph theory the famous theorem of Kuratowski characterizes planar graphs by a minimal list of forbidden substructures. A similar classification could be achieved also for lattices. Unfortunately, here the set $\mathcal{L}$ consisting of seven families contains infinitely many subposets. We will state this result of Kelly and Rival [KR75] here. However, for the list $\mathcal{L}$ we refer to the original paper.

Theorem 3.25 *[KR75] A finite lattice is planar if and only if it does not contain any lattice in $\mathcal{L}$ as a subposet. Moreover, $\mathcal{L}$ is the minimum such list, i.e. any set of lattices $\mathcal{F}$ describing forbidden substructures of planar lattices contains $\mathcal{L}$.*

The very last result highlighted in this section allows an estimation of the number of elements in a planar lattice. It is based on the trivial observation that no planar lattice contains the three-dimensional Boolean algebra $\mathfrak{B}_3$ as a subposet. Several extensions of this observation are known: an approach given in [KR74] excludes a whole family of lattices (containing $\mathfrak{B}_3$) with the help of dismantlability. In [Rep07] an approach by means of n-*distributivity*[14] shows that in every lattice of dimension n every element can be represented as the infimum of at most n $\bigwedge$-irreducibles.

---

[14]This is, a generalization of the well known notion of *distributivity*. A lattice $(\mathfrak{V}, \leq)$ is n-distributive if $x \vee (\bigwedge_{i=0}^{n} y_i) = \bigwedge_{j=0}^{n}(x \vee \bigwedge_{i \neq j, i=0}^{n} y_i)$ holds for all $x, y_0, \ldots, y_n \in \mathfrak{V}$.

Lemma 3.26 *In a planar lattice $\mathfrak{V}$ every element w can be represented as the meet of at most two $\bigwedge$-irreducibles m_1 and m_2 and as the join of at most two $\bigvee$-irreducibles j_1 and j_2.*

Proof: Let $w = \bigwedge \tilde{M}$ for a set $\tilde{M} \subseteq M(w)$ with $|\tilde{M}| \geq 3$. If the width[15] of $(\tilde{M}, \leq)$ does not exceed two then w can be represented as an infimum of at most two elements, i.e. the minimal elements of $\tilde{M}$.

Otherwise there exist pairwise incomparable $\bigwedge$-irreducibles $m_1, m_2, m_3 \in \tilde{M}$. If the meets $m_i \wedge m_j$ are pairwise incomparable for all $i \neq j \in \{1,2,3\}$ then the elements $1_\mathfrak{V}, m_1, m_2, m_3, m_1 \wedge m_2, m_1 \wedge m_3, m_2 \wedge m_3$ and w form a subposet of $\mathfrak{V}$ isomorphic to $\mathfrak{B}_3$ which is contained in the list $\mathcal{L}$ of forbidden subposets of a planar lattice (see Theorem 3.25). Hence, w.l.o.g. $m_1 \wedge m_2 \wedge m_3 = m_1 \wedge m_2$, i.e. $w = \bigwedge \tilde{M} \setminus \{m_3\}$.

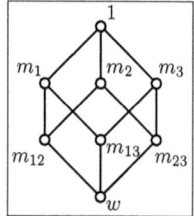

By iteratively applying this argument one finds a two-elemental subset of $\tilde{M}$ whose infimum equals w. The second claim can be shown dually. □

The size of a lattice $\mathfrak{V}$ is exponential by means of the cardinality n of the set of $\bigwedge$-irreducibles $M(\mathfrak{V})$. Every infimum of a subset of $M(\mathfrak{V})$ is a lattice element. In a boolean lattice $\mathfrak{B}_n$ all these infima are distinct, hence it possesses 2^n elements.

Planar lattices have polynomially many elements only: Due to Lemma 3.26 every infimum of a three-elemental subset of $M(\mathfrak{V})$ is equal to the infimum of a smaller subset. Additionally it is possible to construct lattices where indeed all the one- and two-elemental subsets have indeed different infima.

Proposition 3.27 *[Zsc05]*

1. *A finite planar lattice $\mathfrak{V}$ contains at most $\binom{k+1}{2} + 1$ elements, k being the lesser of the cardinalities of $M(\mathfrak{V})$ and $J(\mathfrak{V})$.*

2. *For each $n \in \mathbb{N}$ there exists a planar lattice with $|M(\mathfrak{V})| = n$ possessing $\binom{n+1}{2} + 1$ elements. The dual (replace $M(\mathfrak{V})$ by $J(\mathfrak{V})$) holds as well.*

Proof:

1. W.l.o.g. let $k = |M(\mathfrak{V})|$. By Lemma 3.26, all non-$\bigwedge$-irreducibles except the top element can be described as the meet of exactly two elements of $M(\mathfrak{V})$. This yields

$$|\mathfrak{V}| \leq \binom{k}{2} + k + 1 = \binom{k+1}{2} + 1.$$

---

[15]This is, the maximal cardinality of a subset of pairwise incomparable elements of an ordered set [Bir67].

2. Consider the IIL (see Definition 3.19) $\underline{L}_n := (I_n, \subseteq)$ consisting of the set of intervals (see picture on the right for an example of I_3)

$$I_n := \{i \cap j \mid i, j \in \{[1, n], [2, n+1], \ldots$$
$$\ldots, [n, 2n-1]\}\} \cup \{[0, 2n-1]\}.$$

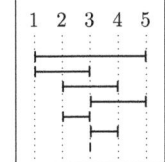

(This lattice is isomorphic to the dual of the *interordinal lattice* [GW99] $\mathfrak{B}(\mathbb{I}_n)^d$, see Figure 4.3.) We easily observe $|I_n| = \binom{n+1}{2} + 1$ and $M(\underline{L}_n) = \{[1, n], [2, n+1], \ldots, [n, 2n-1]\}$. It is easy to see that $\underline{L}_n$ is an IIL, thereby the infimum and supremum of two intervals is equal to their intersection and union respectively. Due to Theorem 3.20 $\underline{L}_n$ is a planar lattice. The dual statement can be derived from the dual lattice of $\underline{L}_n$. □

3.4 Plane Diagrams

In the last section we investigated planarity on a lattice theoretical level, although some proofs relied on the respective diagrams. Now we are going to have a closer look on some properties of those diagrams. Our focus will be moved to a rather geometrical and topological view.

At the beginning we will recall the definition of edge crossings since plane diagrams are determined by their absence.

Definition 3.28 *[GJ83, Grä98]* *Let* $\mathrm{pos}(G)$ *be a diagram of the graph* $G = (V, E)$.

1. *Let e and f be non-incident*[16] *edges of G. Then the diagram edges* $\mathrm{pos}(e)$ *and* $\mathrm{pos}(f)$ *cross in* $\mathrm{pos}(G)$ *if* $\mathrm{pos}(e) \cap \mathrm{pos}(f) \neq \emptyset$.

2. *The number of edge crossings in* $\mathrm{pos}(G)$ *is called* crossing number *[GJ83] (or* complexity *in [Grä98])* $cr(\mathrm{pos}(G))$ *of* $\mathrm{pos}(G)$.

3. *The diagram* $\mathrm{pos}(G)$ *is called* optimal *[Grä98] if the crossing number is minimal, i.e. if no diagram of G with a lesser crossing number exists.* ◊

In graph diagrams it is difficult to describe diagram edges and, in particular, to determine pairs of crossing edges, since they are mapped to arbitrary Jordanarcs. In contrast, the same task is surprisingly easy for diagrams of ordered sets due to the inherent upward-drawing-constraint. As stated in condition 2 of Definition 3.10, the image of an edge can be considered as a continuous function, the so called *corresponding function*:

---
[16]I.e. e and f have no vertex in common.

Chapter 3: Preliminaries

Definition 3.29 *[KR75]* Let $\underline{\mathfrak{V}} = (\mathfrak{V}, \leq)$ be a lattice and $\prec$ its neighborhood relation.

1. A chain C in $\underline{\mathfrak{V}}$ is a sequence $z_0 \prec z_1 \prec \ldots \prec z_n$ of lattice elements z_i. In case of both $z_0 = 0_{\underline{\mathfrak{V}}}$ and $z_n = 1_{\underline{\mathfrak{V}}}$ we call C maximal chain.

2. In a diagram $\text{pos}(\underline{\mathfrak{V}})$ we define the function $x_C : [y(0_{\underline{\mathfrak{V}}}), y(1_{\underline{\mathfrak{V}}})] \mapsto \mathbb{R}$ corresponding to C by $x_C(y) = x_{z_i z_{i+1}}(y)$ if $y \in [y(z_i), y(z_{i+1})]$ holds for all $0 \leq i \leq n-1$. Additionally we define $\text{pos}(C) = \{(x_C(y), y) \mid y \in [y(0_{\underline{\mathfrak{V}}}), y(1_{\underline{\mathfrak{V}}})]\}$. ◊

This means that $\text{pos}(C)$ is just the union of the sets $\text{pos}(z_i z_{i+1})$ as defined in Definition 3.10. The function x_C is a concatenation of the corresponding functions of is edges and therefore continuous itself. Analyzing crossings is very simple in terms of the corresponding functions of the involved edges, as stated in the following lemma.

Lemma 3.30 *[KR75]* Let $\text{pos}(\underline{P})$ be a diagram of an ordered set $\underline{P}$. Let $e_1 = (v_1, w_1)$ and $e_2 = (v_2, w_2)$ be non-incident edges of the graph of $\underline{P}$ and x_{e_1} and x_{e_2} their corresponding functions. Then e_1 and e_2 cross if and only if the following condition holds:

$$(x_{e_1}(y_{max}) - x_{e_2}(y_{max})) \cdot (x_{e_1}(y_{min}) - x_{e_2}(y_{min})) < 0.$$

Thereby $y_{max} := \min\{y(w_1), y(w_2)\}$, the y-coordinate of the lower of the sinks w_1 and w_2 and dually $y_{min} := \max\{y(v_1), y(v_2)\}$, the y-coordinate of the higher of the sources v_1 and v_2 have to satisfy $y_{min} < y_{max}$.

Proof: This is an immediate consequence of the *intermediate value theorem*. □

The picture on the right gives an example of the characterization of crossing edges due to Lemma 3.30. One clearly notices $y_{max} = y(w_2)$ and $y_{min} = y(v_1)$ and further

$$x_{e_1}(y(w_2)) < x_{e_2}(y(w_2)) \quad \text{and}$$
$$x_{e_1}(y(v_1)) > x_{e_2}(y(v_2)).$$

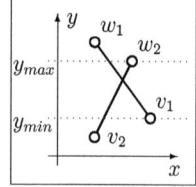

A conjugate order on a lattice can be considered as a relation ordering the elements from left to right. This assertion implies that we refer to a visual, graphical background. Therefore we expect to be able to find a relation that determines whether elements are left or right of each other in a diagram. Now we provide a formal definition satisfying this aim. For a graphical depiction of this left-relation see Figure 3.6.

3.4 Plane Diagrams

Definition 3.31 *[KR75] Let $\mathfrak{V} = (\mathfrak{V}, \leq)$ be a lattice and $\text{pos}(\mathfrak{V})$ a plane diagram of it. Let $\lambda^* \subseteq \mathfrak{V} \times \mathfrak{V}$ be a relation defined by*

$$v \lambda^* w :\iff \exists v^* \in \mathfrak{V} : v, w \prec v^* \text{ and}$$
$$x_{vv^*}(y_{\max}) < x_{wv^*}(y_{\max}),$$

where $y_{\max} := \max\{y(z) \mid z \in \mathfrak{V}, z \prec v^*\}$. Two diagrams are called similar *if their respective λ^* relations are the same.*

The "to the left"-relation $\lambda \subseteq \mathfrak{V} \times \mathfrak{V}$ induced by $\text{pos}(\mathfrak{V})$ is defined by

$$v \lambda w :\iff v \parallel w \text{ and}$$
$$\exists v' \geq v, w' \geq w : v', w' \prec (v \vee w) \text{ with } v' \lambda^* w'.$$

If $v \lambda w$ holds, we say v is left *of w. Dually we define $\varrho := \lambda^{-1}$ and say w is* right *of v.* ◇

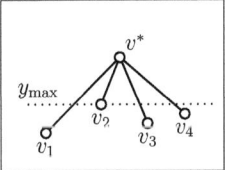
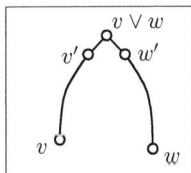

Figure 3.6: On the left one clearly notices $v_1 \lambda^* v_2 \lambda^* v_3 \lambda^* v_4$. All the elements are lower neighbors of v^*. On the right we conclude $v \lambda w$ from $v' \lambda^* w'$, both of the latter elements being lower neighbors of $v \vee w$.

It is shown in [KR75] that $v \lambda w$ is equivalent to the existence of a maximal chain $C \ni w$, where $x(v) < x_C(y(v))$ holds. This helps to prove the following proposition, which actually is the part "1 $\implies$ 2" of Theorem 3.20.

Proposition 3.32 *[KR75] Let $\text{pos}(\mathfrak{V})$ be a plane diagram of the lattice $\mathfrak{V}$. Then the "to the left" relation λ induced by $\text{pos}(\mathfrak{V})$ is a conjugate order.*

Sketch of the proof: From $v \lambda w$ and $w \lambda v$ we can conclude that there are two maximal chains $C \ni v$ and $D \ni w$ such that $\text{pos}(C)$ and $\text{pos}(D)$ intersect "between" (in terms of the y-coordinate) $\text{pos}(v)$ and $\text{pos}(w)$. As $\text{pos}(\mathfrak{V})$ is plane, the intersection point represents a lattice element z fulfilling $v < z < w$ in contradiction to v and w being incomparable. By applying similar arguments we can show λ to be transitive. Hence it is a strict order. Finally, it is easy to see that every two incomparable elements are either left or right of each other according to Definition 3.31, therefore λ is a conjugate relation.

While plane lattice diagrams are quite well researched, the task of *minimizing the number of edge crossings* in the non-planar case remains considerably more difficult. The problem of finding an optimal graph diagram is $\mathcal{NP}$-hard [GJ83]. Freese states in [Fre04] that this result can be modified for lattice diagrams in a straight-forward way. Additionally he proves that even by fixing the ordering of the atoms[17] the problem remains to be $\mathcal{NP}$-complete.

Despite these rather discouraging results we will try to give some hints that allow us to treat "nearly planar" lattices, i.e. those with a small number of edge crossings, in Chapter 7.

3.5 Formal Concept Analysis

The original intention of formal concept analysis (FCA) was to supply a mathematical formalization of "concept" and "concept hierarchy". A comprehensive monograph on FCA is the book of Ganter and Wille [GW99], providing the basic notions and disclosing connections to classical lattice theory.

The investigated basic structures recall of tables or data bases. They consist of objects and attributes and (in the simplest case) a relation declaring the incidence between both sets and are denoted *formal contexts*.

Definition 3.33 *[GW99] A formal context* $\mathbb{K} = (G, M, I)$ *consists of two sets G and M and a relation $I \subseteq G \times M$. The elements of G are called* objects *and the elements of M* attributes. *If gIm holds we say the object g possesses the attribute m.*
$\Diamond$

An example of a formal context, given by a cross table, is depicted in Figure 3.7. The objects, namely bodies of water, are arranged line by line and the attributes in columns. The crosses indicate the attributes of each type of water body. Note that we will omit the term "formal" in the following for convenience reasons and use both context and cross table synonymously.

In order to be able to describe concepts we first need a notion of common attributes of an object set and vice versa. This is done by the *derivation operators*.

Definition 3.34 *[GW99] Let* $\mathbb{K} = (G, M, I)$ *be a context. The* derivation operators[18] *are defined by*

$$A' := \{m \in M \mid gIm \ \forall g \in A\} \text{ and } B' := \{g \in G \mid gIm \ \forall m \in B\}$$

for all subsets $A \subseteq G$ and $B \subseteq M$ respectively.
$\Diamond$

---
[17]An atom of a lattice is an upper cover of the bottom element.
[18]Although two mappings, the derivation operators are commonly denoted by the same symbol for convenience reasons.

3.5 Formal Concept Analysis

Definition 3.35 *[GW99] Let $\mathbb{K} = (G, M, I)$ be a context. The pair (A, B) is a concept if $A \subseteq G$ and $B \subseteq M$ satisfy $A' = B$ and $B' = A$. Then A is the extent and B the intent of the concept. The set of concepts of $\mathbb{K}$ is denoted by $\mathfrak{B}(\mathbb{K})$.* ◊

The set of concepts of a context $\mathbb{K}$ can be ordered canonically by a relation $\leq$ satisfying

$$(A, B) \leq (C, D) : \iff A \subseteq C (\iff B \supseteq D).$$

See the diagram in Figure 3.7 for a depiction of the concept hierarchy of the context *water bodies*. Each node represents a concept (A, B), where A consists of all objects labeled below (A, B) and dually B of all attributes above (A, B). This method is called *reduced labeling* in [GW99].

	natural	large	lentic	lotic
stream	×			×
dam		×	×	
river	×	×		×
lake	×	×	×	

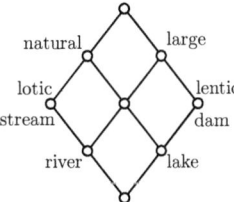

Figure 3.7: A context (left) about water bodies given by a cross table and its corresponding concept lattice (right) provided by a diagram.

It is easy to show that $\mathfrak{B}(\mathbb{K})$ together with the relation $\leq$ forms a complete lattice. This claim is the first part of the following *main theorem on concept lattices*.

Theorem 3.36 *[GW99] Let $\mathbb{K}$ be a context and $\underline{\mathfrak{B}}(\mathbb{K}) := (\mathfrak{B}(\mathbb{K}), \leq)$. Then $\underline{\mathfrak{B}}(\mathbb{K})$ is a complete lattice. The infimum and supremum are determined by*

$$\bigwedge_{t \in T} (A_t, B_t) = \left(\bigcap_{t \in T} A_t, \left(\bigcap_{t \in T} A_t \right)' \right)$$

$$\bigvee_{t \in T} (A_t, B_t) = \left(\left(\bigcap_{t \in T} B_t \right)', \bigcap_{t \in T} B_t \right)$$

A lattice $\underline{\mathfrak{V}} = (V, \leq)$ is isomorphic to $\underline{\mathfrak{B}}(\mathbb{K})$ if there exist mappings $\gamma : G \mapsto V$ and $\mu : M \mapsto V$, s.t. $J(\underline{\mathfrak{V}}) \subseteq \gamma(G)$ and $M(\underline{\mathfrak{V}}) \subseteq \mu(M)$ and the condition $gIm \iff \gamma g \leq \mu m$ holds for all $g \in G$ and $m \in M$.

We conclude from the second part of Theorem 3.36 that $\mathfrak{V} \cong \mathfrak{B}(\mathfrak{V}, \mathfrak{V}, \leq)$ as well as $\mathfrak{V} \cong \mathfrak{B}(J(\mathfrak{V}), M(\mathfrak{V}), \leq)$. The context $(J(\mathfrak{V}), M(\mathfrak{V}), \leq)$ associated to the latter concept lattice will be called *standard context* of $\mathfrak{V}$ [GW99]. It is the smallest possible context representation of a lattice. See Figure 3.8 for an example of a lattice and its standard context. We will use this particular contexts widely in Section 4.4.

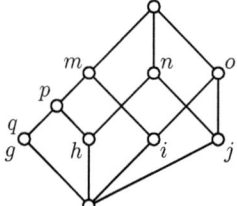

Figure 3.8: A lattice $\mathfrak{V}$ and its standard context. The sets of irreducibles $J(\mathfrak{V}) = \{g, h, i, j\}$ and $M(\mathfrak{V}) = \{m, n, o, p, q\}$ are marked.

3.6 Ferrers-relations

Ferrers-relations were introduced independently by Riguet [Rig51] and Guttman [Gut44] in an attempt to treat qualitative data. More complex relations can be represented as the intersection of Ferrers-relations (see below) leading to the *Ferrers-dimension*. Since a formal context is nothing else than a relation between two sets, we can apply the theory easily to FCA. The connection between the Ferrers- and the order dimension, given by Cogis [Cog82], finally shifts our attention to relations possessing a Ferrers-dimension of at most two.

Definition 3.37 *[Cog82, GW99] A* Ferrers-relation *F is a relation $F \subseteq A \times B$ with*

$$a_1 F b_1 \wedge a_2 F b_2 \quad \Longrightarrow \quad a_1 F b_2 \vee a_2 F b_1.$$

The Ferrers-dimension *$fdim(\mathbb{K})$ of a context $\mathbb{K} = (G, M, I)$ is the smallest number of Ferrers-Relations $F_t \subseteq G \times M$, $t \in T$, whose intersection is equal to I, i.e. $I = \bigcap_{t \in T} F_t$.* ◇

In a cross table representing a context $\mathbb{K}$ we notice that I is a Ferrers-relation if and only if the configuration depicted on the right does not occur.

	m_1	m_2
g_1	×	
g_2		×

3.6 Ferrers-relations

The complement $\overline{F} := (A \times B) \setminus F$ of a Ferrers-relation is again a Ferrers-relation. This brings us to the following observation, which is applied extensively in Section 4.4 since it allows to fill the empty cells of the cross table of a context.

Remark 3.38 *[GW99] The Ferrers-dimension of a context $\mathbb{K} = (G, M, I)$ is the smallest number of Ferrers-relations covering the empty cells of its cross table, i.e. $\overline{I} := (G \times M) \setminus I = \bigcup_{t \in T} F_t$.*

However, the relations F_t $(t \in T)$ are not necessarily disjoint [Reu89, Zsc06b]. It is easy to realize that in a context $\mathbb{K} = (G, M, I)$, I is a Ferrers-relation if and only if $\mathfrak{B}(\mathbb{K})$ is a chain (see [GW99] for a proof). As the order dimension is defined as the intersection of a minimal number of chains, the following theorem seems intuitively reasonable.

Theorem 3.39 [19] *[GW99] Let $\mathbb{K}$ be a context. Then $fdim(\mathbb{K}) = dim(\mathfrak{B}(\mathbb{K}))$.*

We know that a lattice is planar if and only if its order dimension is at most 2 (see Theorem 3.20). Hence the result already gives a useful characterization of contexts possessing planar concept lattices. Although the calculation of the Ferrers-dimension in general is $\mathcal{NP}$-complete [GW99], it is easily treatable in the case we are interested in. For that purpose we introduce the notion of a *Ferrers graph*. Its nodes are the empty cells of a context and its edges indicate which vertices can not belong to the same Ferrers-relation $\overline{F}$. See Figure 3.9 for an example.

Definition 3.40 *[DDF84, Reu89] Let $R \subseteq A \times B$ be a relation. We define the Ferrers-graph $\tilde{\Gamma}(R)$ as an undirected simple graph with vertex set V and edge set E as follows:*

$$V := \overline{R} \qquad E := \{\{(a_1, b_2), (a_2, b_1)\} \mid (a_1, b_1), (a_2, b_2) \in R\}.$$

The bare *Ferrers-graph $\Gamma(R)$ is obtained from $\tilde{\Gamma}(R)$ by deleting all isolated vertices. By the* Ferrers-graph Γ *of a lattice $\mathfrak{V}$ we denote the Ferrers-graph of its standard context $\mathbb{K} = (J, M, \leq)$.* ◊

Let $\chi(\Gamma(I))$ be the chromatic number of $\Gamma(I)$. There is a conjecture claiming that
$$fdim(\mathbb{K}) = r \iff \chi(\Gamma(I)) = r$$

While false in general, Doignon et al. could supply evidence for this conjecture in the special case of $r = 2$.

---
[19]Originally this assertion was given more generally for posets by Cogis in [Cog82].

$\mathbb{K}$	m_1	m_2	m_3	m_4
g_1	×	×	•	•
g_2	•	×	×	•
g_3	•	•	×	×

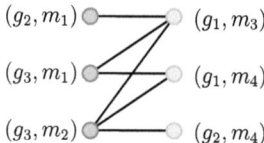

Figure 3.9: A context $\mathbb{K}$ given by a cross table (left) and its corresponding Ferrers-graph $\Gamma(I)$ (right).

Theorem 3.41 *[DDF84] A relation R has the Ferrers-dimension of at most two if and only if its Ferrers-graph is bipartite.*

While the "⇐" part is quite hard to prove, the "⇒" part is obvious, as we will see now:

Lemma 3.42 *[pri]*

1. For a context $\mathbb{K} = (G, M, I)$ the following implication holds:

$$fdim(\mathbb{K}) = 2 \implies \Gamma(I) \text{ is bipartite.}$$

2. The Ferrers-graph $\tilde{\Gamma}$ of the standard context $\mathbb{K}$ of a planar lattice $\mathfrak{V}$ is bipartite.

Proof:

1. Since $fdim(\mathbb{K}) = 2$ there exist two Ferrers-relations F_1 and F_2 with $F_1 \cup F_2 = \bar{I} = V(\Gamma(I))$. Let (g_1, m_1) and (g_2, m_2) be elements of F_1. By Definition 3.37 we notice $g_1 \not I m_2$ or $g_2 \not I m_1$, i.e. $\{(g_1, m_1), (g_2, m_2)\} \notin E(\Gamma(I))$. Analogously we conclude that there exist no edges between elements of F_2. Hence $\Gamma(I)$ is bipartite, e.g. with the vertex classes F_1 and $F_2 \setminus F_1$.

2. If $\mathfrak{V}$ is planar then its order dimension is at most two (Theorem 3.20). Hence, its Ferrers-dimension is at most two (Theorem 3.39). The claim follows with the first assertion of this lemma.
□

In Chapter 6 we will give a constructive proof for the "⇐" part of this theorem. Additionally our framework will allow us to specify all representations of a relation R as the intersection of two Ferrers-relations, i.e. all plane diagrams of a lattice up to similarity.

Chapter 4

Left-relations on Lattices

In the last section we introduced conjugate orders as a way to characterize planar lattices. Understanding an ordered set with that additional order is surprisingly intuitive. The standard order can be considered as an above-below-relationship, while the conjugate order supplies an additional left-right-sorting of elements. From everyday experience it is obvious that these two relations are orders and, moreover, that (according to Definition 3.17) two elements are situated in exactly one of them.

There already exist relations in lattice theory called *"to the left of"* [KR75]. However, these are used not for lattices itself but rather for their representing diagrams. Here we want to introduce *left-relations* in a more lattice-theoretical approach. Of course we could just employ conjugate orders for that purpose since they do the job.

However, our approach has two advantages. On one hand we construct left-relations out of a relation on a smaller set, namely the set of $\bigwedge$-irreducibles. This allows us to understand conjugate orders in a better way (see Propositions 4.14 and 4.18) and to introduce and investigate left-relations on contexts (see Definition 6.1 and Lemmas 6.4 and 6.5). Moreover, by understanding a left-relation as a unique modified extension of a linear order on the set of $\bigwedge$-irreducible elements, we are even able to characterize *planar contexts* (see Section 4.4). This is, we give a criterion concerning the shape of the standard context (or its representing cross table) of a lattice assuring that lattice to be planar.

On the other hand we are given a possibility to also concern non-planar lattices. Indeed, in that case the left-relation is not an order, contradicting the common sense. However, by "measuring" how much the property of being an order is violated, one may give assertions beyond planarity. Unfortunately we do not have the room to extensively discuss this fairly interesting topic in this work and will only give some hints in Chapter 7.

4.1 Definition and Basic Properties

The idea of left-relations was extracted from the convention of attribute additive diagrams (Definition 3.14). As we already mentioned, such a drawing is determined completely by the vectors assigned to the attributes (or $\bigwedge$-irreducibles respectively). This gives rise to the assumption that just the relationship of the representing diagram nodes in the Euclidian plane will cause the diagram to be plane or not. In fact, we only need a relation indicating for two nodes v and w with common upper neighbor whether v is left of w or otherwise w is left of v. As we already emphasized, we will shift this geometric consideration to a lattice theoretical one. Thus, we begin by defining a *sorting relation*.

Definition 4.1 *[Zsc05] Let $\mathfrak{V}$ be a finite lattice and M be the set of its $\bigwedge$-irreducible elements. A strict order $S \subseteq M \times M$ is called* sorting relation *if the condition*

$$m^* = n^* \iff m \: S \: n \text{ or } n \: S \: m$$

holds for all elements $m, n \in M$ (by m^* we denote the unique upper neighbor of an $\bigwedge$-irreducible m). ◊

We notice that S is a union of linear orders, each acting on the set of all $\bigwedge$-irreducibles sharing a common upper neighbor. Based on it we can uniquely derive a *left-relation on a lattice* in an iterative way from top to bottom, as we will construct now.

For that issue we need the denotation of the set

$$M(v,w) := \{(m,n) \subseteq M \times M \mid \\ v \leq m, w \leq n, v \parallel n, w \parallel m\}. \quad (4.1)$$

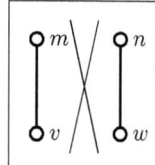

of all pairs of $\bigwedge$-irreducibles (m, n), where m and n are "above" v and w respectively and m and w and n and v respectively are incomparable (see picture on the right).

Definition 4.2 *[Zsc05] Let $\mathfrak{V}$ be a finite lattice and S be a sorting relation on $\mathfrak{V}$. A binary relation $L \subseteq \mathfrak{V} \times \mathfrak{V}$ defined by*

$$v \: L \: w :\iff \begin{cases} v \: S \: w, & v, w \in M, v^* = w^* \\ \exists (m,n) \in M(v,w) : m \: L \: n, & \text{else} \end{cases} \quad (4.2)$$

is called left-relation *(induced by S) and the relation $R := L^{-1}$ is called* right-relation *on the lattice $\mathfrak{V}$.* ◊

4.1 Definition and Basic Properties

In Figure 4.1 we see an example of how a left-relation is constructed due to its definition. Notice that we are interested not in the particular diagram but in the underlying lattice structure. The given lattice has three $\bigwedge$-irreducibles m_1, m_2 and m_3, the top and the bottom element (which are not taken into consideration, since they are comparable to all other lattice elements) and the element v_1. In the left picture we see a given sorting relation, namely $m_1 \ S \ m_2$. The $\bigwedge$-irreducible m_3 is the only one possessing the upper neighbor m_1. It is therefore not in relation S to any other element. We conclude $m_1 \ L \ m_2$. In the second picture we consider m_2 and m_3 and find $m_3 \ L \ m_2$ since we have $M(m_3, m_2) = \{(m_1, m_2), (m_3, m_2)\}$. In the right picture we finally find $(m_3, m_2) \in M(m_3, v_1)$ and consequently $m_3 \ L \ v_1$. So we assigned the left-relation properly; all other pairs of lattice elements are comparable.

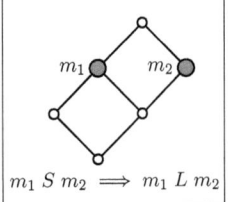 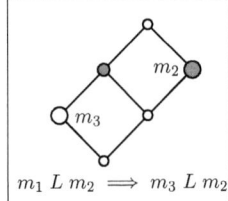 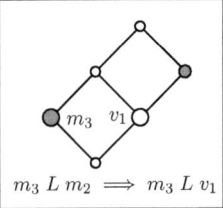

Figure 4.1: An illustration of the construction of a left-relation out of a sorting relation in a small lattice. The bigger dots symbolize the two elements just in consideration. They are caused to be in left-relation by the shaded nodes.

Definition 4.2 somewhat seems to be formulated clumsily. One can ask why L is defined in an iterative way rather than directly out of the sorting relation S. This could be done for instance by

$$v \ L \ w : \iff \begin{cases} v \ S \ w, & v, w \in M, v^* = w^* \\ \exists (m, n) \in M(v, w) : m \ S \ n, & \text{else} \end{cases} \quad (4.3)$$

However this simpler approach fails sometimes to meet our desire to have every pair of incomparable lattice elements v and w in the left-relation, i.e. either $v \ L \ w$ or $w \ L \ v$ shall hold. Consider the lattice diagram depicted on the right. We encounter $M(v, w) = \{(m_3, m_4)\}$ but only m_1 and m_2 are in sorting relation (for instance by $m_1 \ L \ m_2$). Hence, Property (4.3) would not provide us with an information whether v is left of w or vice versa.

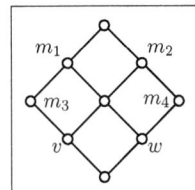

When we designed the left-relation, we had certain ideas in mind about its properties. In particular a left-relation may act exactly on pairs of incomparable elements, i.e. it is supposed to be a conjugate relation (see Definition 3.17). We gave an idea that a "simpler" definition may not meet that requirement, but we did not show that the construction according to Definition 4.2 does. Before doing so, we will give some characteristics of the set $M(v,w)$ which follow directly from its definition (see Equation (4.1)).

Remark 4.3 *[Zsc05] Let $(\mathfrak{V}, \leq)$ be a finite lattice and M the set of its $\bigwedge$-irreducible elements. The following properties hold for all $v, w \in \mathfrak{V}$ and $m, n \in M$:*

1. $(m, n) \in M(v, w) \implies m \parallel n$,

2. $M(v, w) = \emptyset \iff v \leq w$ or $w \leq v$,

3. $M(m, n) = \{(m, n)\} \iff m^* = n^*$,

4. $(m, n) \in M(v, w) \implies (m, n) \geq (v, w) \quad (: \iff m \geq v$ and $n \geq w)$.

5. $(m, n) \in M(v, w) \iff (n, m) \in M(w, v)$.

Lemma 4.4 *[Zsc05] Let L be a left-relation on a lattice $\mathfrak{V}$, then L is a conjugate relation.*

Proof: Comparable pairs of elements v and w are obviously not in L since they are not in sorting relation and the set $M(v,w)$ is empty (see Remark 4.3, 2.).

Now let m and n be incomparable $\bigwedge$-irreducible elements. Assume that neither $m \, L \, n$ nor $n \, L \, m$ holds and that m, n is a maximal pair[1] with that property. According to Definition 4.2, the pair (m, n) is no element of the underlying sorting relation. Due to Remark 4.3, 3. we know that there exists a pair $(\hat{m}, \hat{n})$ meeting $(m,n) \leq (\hat{m}, \hat{n}) \in M(m,n)$. Since (m,n) is maximal we know that $(\tilde{m}, \tilde{n}) \in L \cup R$ and hence $(m,n) \in L \cup R$ contradicting our assumption.

Finally let v and w be arbitrary incomparable lattice elements. Then there exists a pair $(m,n) \in M(v,w)$. In the last paragraph we have shown that m and n are in left-relation, hence v and w are too. □

We want to remark another two obvious facts. Firstly, if the lattice $\mathfrak{V}$ is not a linear order then there exists a non-empty sorting relation S which induces a left-relation L. Then S^{-1} is a sorting relation different from S inducing $L^{-1} = R$.

Secondly, two different left-relations L and $\tilde{L}$ are induced by different sorting relations S and $\tilde{S}$: Since both left-relations are conjugate relations, the

---

[1] I.e. for all pairs $(\tilde{m}, \tilde{n}) \geq (m, n)$ with $\tilde{m} \neq m$ or $\tilde{n} \neq n$ we find $\tilde{m} \, L \, \tilde{n}$ or $\tilde{n} \, L \, \tilde{m}$.

4.1 Definition and Basic Properties

assertion $L \neq \tilde{L}$ implies the existence of a maximal pair (v, w) of lattice elements meeting $v \: L \: w$ and $v \: \tilde{\not{L}} \: w$. The pair is maximal, therefore we conclude $M(v, w) = \{(v, w)\}$ and hence $v \: S \: w$ and $v \: \tilde{\not{S}} \: w$. We recapitulate:

Remark 4.5 *Let $\mathfrak{V}$ be a lattice. The following assertions hold:*

1. *Every right-relation on $\mathfrak{V}$ is a left-relation on $\mathfrak{V}$.*

2. *Every sorting relation S on $\mathfrak{V}$ uniquely induces a left-relation L.*

In Figure 4.2 we investigate another example of a left-relation. We consider the lattice depicted by a diagram on the left. There exist two pairs of $\bigwedge$-irreducibles with common upper neighbor, namely (a, b) and (c, d). We assume $a \: S \: b$ and $c \: S \: d$ for the sorting relation. In a first step we calculate the left-relation on pairs of incomparable $\bigwedge$-irreducibles. This yields $a \: L \: b$, $c \: L \: d$, $a \: L \: c$ and $a \: L \: d$. Finally we determine the left-relation on all other incomparable pairs and obtain the table shown on the right. We notice that L is indeed a conjugate relation; hence $L \cup <$ is connex. However, L is no order since we find $c \: L \: x \: L \: c$.

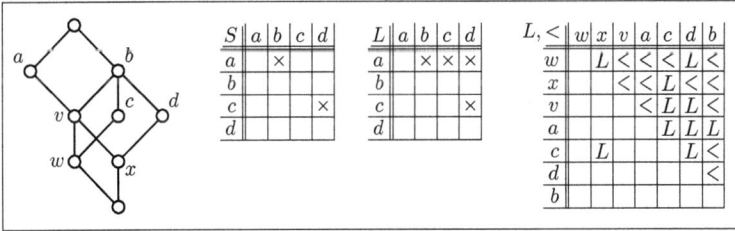

Figure 4.2: A lattice (given on the left) with a sorting relation S, its induced left-relation L restricted to $\bigwedge$-irreducibles and a combination of L and the lesser-than-relation on the right.

As a last preparation for the following section, where we consider left-relations that are additionally strict orders, we introduce a notion for those structures:

Definition 4.6 *A left-relation L is called* left-order *if it is a strict order.* ◊

4.2 Left-relations and Conjugate Orders

In the first part of this chapter we have explained that our attempt to develop left-relations is directed by the idea to gain a useful description of conjugate relations. In this section we want to clarify the strong correlation between these two concepts. We will finally conclude that every conjugate order can be represented by a left-order and that conversely all left-orders are conjugate orders.

Our first lemma claims that the construction principle of left-relations is inherent in conjugate orders, too.

Lemma 4.7 *[Zsc05] Let $\mathfrak{V}$ be a finite lattice and L_c be a conjugate order on $\mathfrak{V}$. The following implication holds for all lattice elements v, w and $\bigwedge$-irreducibles m and n:*

$$m \, L_c \, n \text{ and } (m, n) \in M(v, w) \implies v \, L_c \, w$$

Proof:

Since $M(v, w) \neq \emptyset$, we know (see Remark 4.3, 2.) $v \parallel w$. With Definition 3.17 we conclude either $v \, L_c \, w$ or $w \, L_c \, v$. Let us assume $w \, L_c \, v$. Since (m, n) is an element of $M(v, w)$ we notice that m and w are incomparable. Therefore either $m \, L_c \, w$ or $w \, L_c \, m$ holds.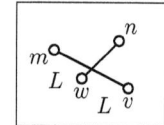
In the first case we find $m \, L_c \, w \, L_c \, v$ (see picture) and in the latter $w \, L_c \, m \, L_c \, n$. Since L_c is a transitive relation we conclude $m \, L_c \, v$ and $w \, L_c \, n$ respectively. This contradicts L_c to be a conjugate relation since we premised $v \leq m$ and $w \leq n$ respectively. We conclude $v \, L_c \, w$. □

This was already the key step towards our goal. A conjugate order L_c obviously includes a sorting relation S. Since L_c can be generated in the same way as the appropriate left-relation L induced by S, both are equal. This is the subject of the next lemma.

Lemma 4.8 *[Zsc05] A conjugate order L_c on a finite lattice $\mathfrak{V}$ is a left-relation on $\mathfrak{V}$.*

Proof: Let L_c be a conjugate order on $\mathfrak{V}$. For every two incomparable $\bigwedge$-irreducibles $m, n \in M$ either $m \, L_c \, n$ or $n \, L_c \, m$ holds. Hence there exists a sorting relation $S \subseteq L_c$.

Let L be the left-relation generated by S. We assume $L \neq L_c$. In this case we find a maximal pair of lattice elements (v, w) with $v \, L \, w$ and $w \, L_c \, v$. On one hand we find (by applying Definition 4.2) a pair $(m, n) \in M(v, w)$ with

4.2 Left-relations and Conjugate Orders

$m\ L\ n$. On the other hand we know (by applying Lemma 4.7) that $n\ L_c\ m$. This contradicts the assumption of the maximality of (v,w) since we know $(m,n) > (v,w)$ (see Remark 4.3 4). □

The achieved results can be summarized to the following theorem.

Theorem 4.9 *[Zsc05] Let L be a relation on a finite lattice $\mathfrak{V}$. Then the following are equivalent:*

1. *L is a conjugate order.*

2. *L is a left-order.*

Proof: "1. ⇒ 2." follows from Lemma 4.8.
"2. ⇒ 1." follows from Lemma 4.4 and Definition 3.17. □

Theorem 4.9 simplifies the handling of conjugate orders. In general, a lattice $\mathfrak{V} = (\mathfrak{V}, \leq)$ may have up to $2^{\frac{k^2-k}{2}}$ (with $k := |\mathfrak{V} - 2|$) asymmetric conjugate relations as candidates for conjugate orders, but at most $|M(\mathfrak{V})|!$ left-relations as candidates for left-orders.

Consider the lattice M_5 (see picture on the right) as an example. It possesses $5! = 120$ left-relations, namely all permutations of the coatoms[2] a, b, c, d and e. On the other hand, if we consider the relation table of that lattice, we notice that a conjugate relation is a subset of the incomparability relation (consisting of the $(7-2)^2 - (7-2) = 20$ empty fields in the table). The union of a conjugate order and its complement must cover these empty fields. Additionally, the meet with its complement equals the empty set. Hence, there exist $2^{10} = 1024$ candidates for conjugate orders.

Of course, the number of left-relations is not polynomial in terms of the number of lattice elements. Hence, calculating left-orders naively by checking all possible left-relations to be strict orders will fail soon in time complexity.

In Figure 4.3 we can realize the advantages and disadvantages of the attempt to calculate conjugate orders of a lattice out of its left-relations. The upper frame shows the *interordinal lattice* [GW99] $\mathfrak{B}(\mathbb{I}_4)$. There exist just the two sorting relations $S_1 = \{(a,d)\}$ and $S_2 = \{(d,a)\}$. It is easy to show that both of them result in a left-order (notice that $L_1^{-1} = L_2$). The picture below shows that things become complicated only by flipping the diagram upside down and

---
[2] Coatoms are the lower covers of the top element of a lattice.

considering the dual lattice. Here we find that the four $\bigwedge$-irreducibles are all coatoms. Therefore there exist $4! = 24$ sorting relations. Testing all of whom for inducing a left-order is quite tedious.

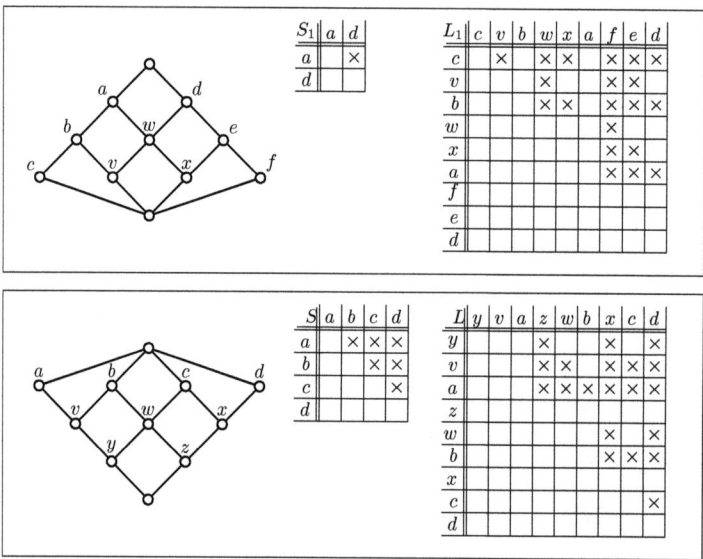

Figure 4.3: The interordinal lattice (upper picture) possesses just two sorting relations, both supplying a left-order. In the dual interordinal lattice (lower picture) however, we find 24 sorting relations. Only two of them induce a left-order.

4.3 Planarity Conditions

In the last section we have shown left-orders and conjugate orders to be equivalent. When we are given a sorting relation S on a lattice $\mathfrak{V}$, we construct the induced left-relation L. If L turns out to be a strict order then $\mathfrak{V}$ is planar.

The particular structure of left-relations allows us to characterize planarity even without considering it on the whole lattice. In fact we just need to observe L on all pairs of incomparable $\bigwedge$-irreducible elements. This may improve the testing process. The main contribution however is to ease proofs, in particular in Section 4.4 where we will be searching for conditions of standard

4.3 Planarity Conditions 43

contexts making its appropriate lattice planar. Indeed the process of calculating all conjugate orders of a lattice can not be significantly simplified with those techniques since we still have to consider all sorting relations.

Before we can give those results, we need some preparations. First we will show that an asymmetric left-relation can be "proceeded" up and down.

Lemma 4.10 *[Zsc06a] Let $\mathfrak{V} = (\mathfrak{V}, \leq)$ be a finite lattice and L an asymmetric left-relation on $\mathfrak{V}$. Let the lattice elements $v_1, v_2, v_3 \in \mathfrak{V}$ meet the requirements $v_1 \parallel v_3 \parallel v_2$ and $v_1 \leq v_2$. Then*

$$v_1 \, L \, v_3 \iff v_2 \, L \, v_3.$$

Proof: There exists a $\bigwedge$-irreducible m_2 satisfying both $m_2 \geq v_2$ and $m_2 \not\geq v_3$ since v_2 and v_3 are incomparable. Analogously we find a $\bigwedge$-irreducible m_3 such that the conditions $m_3 \geq v_3$ and $m_3 \not\geq v_1$ hold. We notice $(m_2, m_3) \in M(v_1, v_3) \cap M(v_2, v_3)$. Since L is asymmetric, we conclude $v_1 \, L \, v_3 \iff m_2 \, L \, m_3 \iff v_2 \, L \, v_3$, which is the claim. □

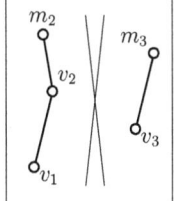

Lemma 4.10 is a useful tool being employed several times. As a first consequence we give the following assertion.

Corollary 4.11 *[Zsc06a] Let $\mathfrak{V} = (\mathfrak{V}, \leq)$ be a finite lattice and L an asymmetric left-relation on $\mathfrak{V}$. Let v_1, v_2, v_3 be lattice elements such that $v_1 \, L \, v_2 \, L \, v_3$ holds. Then v_1 and v_3 are incomparable.*

Proof: We assume v_1 and v_3 to be comparable. By applying Lemma 4.10 we notice $v_1 \, L \, v_2 \implies v_3 \, L \, v_2$. This contradicts the fact that L is asymmetric. □

Our next result is quite surprising. When we want to show a left-relation to be an order, we have to evidence its asymmetry and transitivity. Here we can prove that the first property already implies the second.

Lemma 4.12 *[Zsc06a] Let $\mathfrak{V} = (\mathfrak{V}, \leq)$ be a finite lattice and L an asymmetric left-relation on $\mathfrak{V}$. Then L is transitive.*

Proof: Let v_1, v_2 and v_3 be arbitrary lattice elements fulfilling $v_1 \, L \, v_2 \, L \, v_3$. With Corollary 4.11 we know $v_1 \parallel v_3$. We want to show $v_1 \, L \, v_3$.

1. Let $v_{12} := (v_1 \vee v_2) \parallel v_3$. Since we presumed $v_2 \, L \, v_3$, we conclude with Lemma 4.10 $v_{12} \, L \, v_3$ and $v_1 \, L \, v_3$. By an analog argumentation we find

$$v_{23} := (v_2 \vee v_3) \parallel v_1 \implies v_1 \, L \, v_3.$$

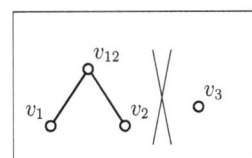

2. Let otherwise $v_{123} := (v_1 \vee v_2 \vee v_3) = v_{12} = v_{23}$. If we find $\bigwedge$-irreducibles $m_1 \geq v_1$, $m_2 \geq v_2$ and $m_3 \geq v_3$ then we conclude with Lemma 4.10 $m_1\ S\ m_2\ S\ m_3$. Due to the asymmetry of S we find $m_1\ S\ m_3$ and therefore $v_1\ L\ v_3$. On the other hand there exists an element v_4 fulfilling $v_4 \parallel v_{123}$ and either $v_4 > v_1$ or $v_4 > v_2$ or $v_4 > v_3$.

In case of $v_4 > v_1$ we notice $v_2 \parallel v_4 \parallel v_3$. With Lemma 4.10 follows

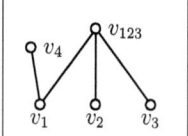

$$v_1\ L\ v_2 \implies v_4\ L\ v_2 \implies v_4\ L\ v_{123}$$
$$\implies v_4\ L\ v_3 \implies v_1\ L\ v_3.$$

An analog argumentation can be applied if there exists an element v_4 satisfying $v_4 \parallel v_{123}$ and either $v_4 > v_2$ or $v_4 > v_3$. □

After this preliminary work we will present the two main results of this section, the so called *planarity conditions*. Both describe constraints of a left-relation which are both necessary and sufficient for the planarity of the underlying lattice. The first can be considered as a way to indicate which $\bigwedge$-irreducibles are "in-between" others.

Definition 4.13 *[Zsc06a] A conjugate relation R on a lattice $\mathfrak{V}$ fulfills the first planarity condition (FPC) if*

$$m_1\ R\ n\ R\ m_2 \implies n > (m_1 \wedge m_2)$$

holds for all $\bigwedge$-irreducibles $m_1, m_2, n \in M$. ◇

See Figure 4.4 for an intuitive understanding of the necessity of the FPC.

Proposition 4.14 *[Zsc06a] Let L be a left-relation on a lattice $\mathfrak{V}$, then the following equivalence holds:*

$$L \text{ satisfies the FPC} \iff L \text{ is a conjugate order.}$$

Proof:
"$\Rightarrow$": We assume L not to be asymmetric. Then we find two lattice elements v, w being maximal with the property $v\ L\ w\ L\ v$. By Definition 4.2 there exist two pairs $(m_1, n_1), (m_2, n_2) \in M(v, w)$ with $m_1\ L\ n_1$ and $n_2\ L\ m_2$. Let w.l.o.g. $m_1 \neq m_2$. Moreover we know $m_1 \parallel n_2$, i.e. either $m_1\ L\ n_2$ or $n_2\ L\ m_1$ holds. The first case (see picture) leads to $m_1\ L\ n_2\ L\ m_2$. By the presumed FPC we conclude $n_2 > (m_1 \wedge m_2) \geq v_1$ which contradicts $v_1 \parallel n_2$. The latter implies $n_2\ L\ m_1\ L\ n_1$, i.e. $m_1 > v_2$, again a contradiction.

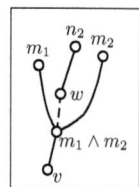

4.3 Planarity Conditions

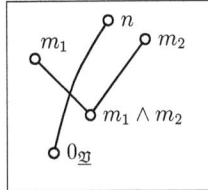 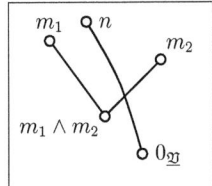

Figure 4.4: When considering a lattice diagram, the necessity of the FPC for its planarity is intuitively clear. If $m_1 \ L \ n \ L \ m_2$ or $m_2 \ L \ n \ L \ m_1$ holds, so does $n > (m_1 \wedge m_2)$. Otherwise every image of a chain from n to the bottom element of the lattice intersects with appropriate images of chains from either m_1 or m_2 to $m_1 \wedge m_2$. *Left-relations on diagrams* will be discussed in Chapter 5.

Therefore L is asymmetric. With Lemma 4.12 we know that L is a strict order, i.e. a conjugate order and hence (see Theorem 3.20) $\mathfrak{V}$ is planar.

"$\Leftarrow$": A planar lattice $\mathfrak{V}$ possesses an asymmetric left-relation L. Let m_1, m_2 and n be arbitrary pairwise incomparable $\bigwedge$-irreducibles satisfying $m_1 \ L \ n \ L \ m_2$. Consider $v := (m_1 \wedge m_2)$. We notice that neither $v \ L \ n$ (this implies with Lemma 4.10 $m_2 \ L \ n$) nor $n \ L \ v$ (this implies $n \ L \ m_1$) nor $v \geq n$ (this implies $m_1 > n$) holds (see Figure 4.4). Therefore $v < n$. □

As hinted already, the formulation of the FPC allows to specify which lattice elements may be situated "in-between" others in a plane lattice diagram. One may have the idea of introducing a ternary relation $T \subseteq M \times M \times M$ by

$$(m_1, m_2, m_3) \in T : \iff m_2 > (m_1 \wedge m_3) \text{ and } m_1 \parallel m_2 \parallel m_3.$$

Such an element of T is read "m_2 is permitted to be situated between m_1 and m_3". Due to Proposition 4.14, a left-relation L is a strict order if for all $m_1, m_2, m_3 \in M$ the implication $m_1 \ L \ m_2 \ L \ m_3 \implies (m_1, m_2, m_3) \in T$ holds. It seems to be possible to construct a left-order L, if it exists at all, out of T by consecutively attaching the elements of T (considered as three-elemental chains) until one gets the sequence of a linear extension of L. However, we could not yet succeed in creating an algorithm performing that task.

While the FPC describes planar lattices in terms of in-betweenness of $\bigwedge$-irreducibles in a respective left-relation, the *Second Planarity Condition* provides a characterization by clustering them. In Figure 4.5 we give an intuitive explanation. Beforehand, we need to introduce some notations.

For two sets $\tilde{A} \subseteq A$ and $\tilde{B} \subseteq B$ and a relation $R \subseteq A \times B$ we write $\tilde{A} \ R \ \tilde{B}$ if $\tilde{A} \times \tilde{B} \subseteq R$, i.e. if each element of $\tilde{A}$ is in relation to each element of $\tilde{B}$.

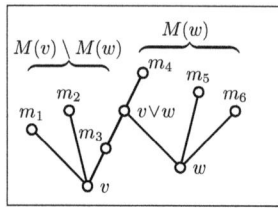 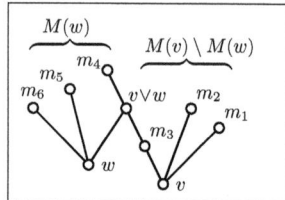

Figure 4.5: The SPC holds, if for each two incomparable lattice elements v and w the $\bigwedge$-irreducibles above w are either right or greater *or* left or greater than the ones which are above v but not above w.

Moreover, we recall $R_<:=R\cup <$ and $R_>:=R\cup >$ respectively.

Definition 4.15 *[Zsc06a] A conjugate relation R on a lattice $\mathfrak{V}$ fulfills the second planarity condition (SPC) if the two below-stated requirements are satisfied:*

1. *R is asymmetric on $M \times M$.*

2. *For all lattice elements $v \parallel w \in \mathfrak{V}$ holds*

$$(M(v) \setminus M(w))\ R_<\ M(w)\ \text{or}\ M(w)\ R_>\ (M(v) \setminus M(w)).$$
◇

Lemma 4.16 *[Zsc06a] Let L be a left-relation on a lattice $\mathfrak{V}$. Then the following equivalence holds:*

$$L\ \text{satisfies the SPC}\ \iff L\ \text{is a conjugate order.}$$

Proof:
"$\Rightarrow$": We prove that the SPC implies the FPC: Let m_1, m_2 and n be elements of M satisfying $m_1\ L\ n\ L\ m_2$. Furthermore let $v := m_1 \wedge m_2$. Obviously $v \not\geq n$. If we assume $n \parallel v$, we note $m_1, m_2 \in (M(v) \setminus M(n))$. Since $n \in M(n)$ we conclude

$$m_1\ L\ n\ \implies (M(v) \setminus M(n))\ L_<\ M(n)\ \implies m_2\ L\ n,$$

contradicting the asymmetry of L. Therefore $v < n$.
"$\Leftarrow$": Since L is a conjugate order it is a strict order on $M \times M$. Let v and w be arbitrary lattice elements. Let $m \in M(v) \setminus M(w)$ and $n \in M(w)$ be incomparable $\bigwedge$-irreducibles . In case of $m\ L\ n$ we conclude with Lemma 4.10

$$m\ L\ n\ \implies m\ L\ w\ \implies v\ L\ w$$
$$\implies M(v) \setminus M(w)\ L\ \{w\}\ \implies M(v) \setminus M(w)\ L_<\ M(w).$$

4.3 Planarity Conditions

The other case $n L m$ can be handled in an analog way. □

Due to Lemma 4.16, it is possible to decompose the set of $\bigwedge$-irreducibles above two incomparable lattice elements $v \parallel w$ into three parts w.r.t. a left-order: we have either

$$M(v) \setminus M(w) \; L_< \; M(v \vee w) \; L_> \; M(w) \setminus M(v) \text{ or vice versa}$$
$$M(w) \setminus M(v) \; L_< \; M(v \vee w) \; L_> \; M(v) \setminus M(w).$$

By verifying the SPC on every pair of incomparable lattice elements, one gets a bundle of clusters of $\bigwedge$-irreducible elements. A left-relation L is a left-order, if in a linear extension of L the elements of each cluster are sorted consecutively. However, we failed to design an appropriate algorithm to create left-orders out of the clusters provided by the SPC as well.

The SPC can be formulated in a more efficient way. Instead of clustering the $\bigwedge$-irreducibles above all lattice elements it is enough to consider the $\bigvee$-irreducibles. This is sufficient to characterize planarity.

Definition 4.17 *[Zsc06a] A conjugate relation R on a lattice $\mathfrak{V}$ fulfills the reduced second planarity condition (rSPC) if the subsequent requirements are satisfied:*

1. *R is asymmetric on $M \times M$.*

2. *For all $\bigvee$ irreducibles $g_1 \parallel g_2 \in \mathfrak{V}$ holds*

$$(M(g_1) \setminus M(g_2)) \; R_< \; M(g_2) \text{ or } M(g_2) \; R_> \; (M(g_1) \setminus M(g_2)).$$ ◊

Proposition 4.18 *[Zsc06a] Let L be a left-relation on a lattice $\mathfrak{V}$, then the following equivalence holds:*

$$L \text{ satisfies the rSPC} \iff L \text{ is a conjugate order.}$$

Proof:
"$\Rightarrow$": Similarly to the last proof we show that the rSPC implies the FPC. Let m_1, m_2 and n be $\bigwedge$-irreducibles fulfilling $m_1 \; L \; n \; L \; m_2$. Define $v := m_1 \wedge m_2$. Clearly $v \not\geq n$ since $n \parallel m_1$.

Let us assume $v \parallel n$. Then we find $\bigvee$-irreducibles g and h satisfying $g \leq v$, $h \leq n$ and $n \parallel g \parallel h$ (see picture). With the rSPC we find

$$n \; L \; m_1 \implies (M(h) \setminus M(g)) \; L_< \; M(g) \implies n \; L \; m_2$$

contradicting L to be asymmetric. Therefore $v < n$.

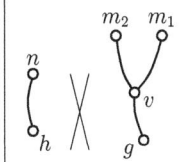

"$\Leftarrow$": If L is a conjugate order, it fulfills the SPC (see Lemma 4.16) and hence the rSPC, too. □

Although the FPC and rSPC give helpful characterizations of the planarity of a lattice by conditions referring to their $\bigwedge$-irreducible elements, we were not able to use them algorithmically to find plane diagrams efficiently. We will come back to this issue in Chapter 6.

Nevertheless, at least the FPC will be of extensive use in our subsequent considerations. Deciding whether a lattice is planar only from the interrelation of the set M of $\bigwedge$-irreducibles saves a lot of work. In particular, we will use the FPC in the next section to observe whether there exists a linear order on M that induces a left-order on the corresponding lattice. We will describe this linear order by an enumeration (to be understood as an indexing) of the elements of m. Such a linear order usually appears in a context or, more precisely, a cross table. This naturally leads to the question whether we can characterize the planarity of a lattice by its standard context.

4.4 Planar Contexts

The concept of a *planar context* may be misleading. Of course, by this notation we do not mean a context that can be drawn in the plane, but the standard context of a planar lattice. We chose that name for formulating this topic in a concise way.

In this section we want to clarify how to recognize "at a glance" from a standard context $\mathbb{K} = (J(\mathfrak{V}), M(\mathfrak{V}), \leq)$ whether the appropriate lattice $\mathfrak{V}$ is planar. Thereby, we rely on the so-called *consecutive-one property* introduced by Fulkerson and Gross in [FG65]. A relaxed version of that property (which we will call *third planarity condition*) necessarily and sufficiently allows the desired statement.

The consecutive-one property is investigated particularly in graph theory. A matrix $M \in \{0,1\}^{m \times n}$ (e.g. an adjacency matrix) satisfies this property if its columns can be permuted, s.t. in each row all entries containing the number 1 are sorted consecutively. We apply this intuition to our purpose:

Definition 4.19 *[FG65,BL76] A context $\mathbb{K} = (G, M, I)$ fulfills the consecutive-one property if there exists an enumeration of the attributes, i.e. a bijective mapping $\varepsilon : M \to [1, \ldots, |M|]$, s.t. the condition*

$$gIm_i \wedge gIm_k \implies gIm_j$$

holds for all objects $g \in G$ and for all attributes $m_i, m_j, m_k \in M$ satisfying $1 \leq \varepsilon(m_i) < \varepsilon(m_j) < \varepsilon(m_k) \leq |M|$. ◊

4.4 Planar Contexts

If a context $\mathbb{K} = (G, M, I)$ fulfills the consecutive-one property then we can decompose the Cartesian product $G \times M$ into three factors F_1, F_2 and I by

$$F_1 := \{(g,m) \mid \forall n \in M : \varepsilon(n) \leq \varepsilon(m) \Rightarrow g \not I n\},$$
$$F_2 := G \times M \setminus (F_1 \cup I).$$

In a cross table, the sets F_1, I and F_2 can be easily encountered by sorting the attributes according to ε (see picture on the right). The set F_1 consists of all empty cells left of any cross and F_2 of the empty cells right of any cross. As the picture suggests, both F_1 and F_2 are Ferrers-relations. That implies that $\mathbb{K} = (G, M, I)$ is planar, as we will show next.

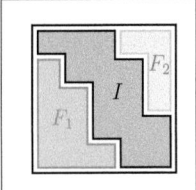

Lemma 4.20 *Let $\mathbb{K} = (J(\mathfrak{V}), M(\mathfrak{V}), \leq)$ be the standard context of a lattice $\mathfrak{V}$. If $\mathbb{K}$ fulfills the consecutive-one property then $\mathfrak{V}$ is planar.*

Proof: Let F_1 and F_2 be defined as above. Let $(j_1, m_1), (j_2, m_2) \in F_1$ with $\varepsilon(m_1) \leq \varepsilon(m_2)$ then also $(j_2, m_1) \in F_1$. Hence F_1 is a Ferrers-relation.

Let $(j_1, m_1), (j_2, m_2) \in F_2$ with $\varepsilon(m_1) \leq \varepsilon(m_2)$. There exists a $\bigwedge$-irreducible m satisfying $\varepsilon(m) < \varepsilon(m_1)$ and $j_1 I m$, otherwise we had $(j_1, m_1) \in F_1$. Since $\mathbb{K}$ fulfills the consecutive-one property we conclude $(j_1, m_2) \in F_2$.

$\mathbb{K}$	m	m_1	m_2
j_1	×	□	
j_2			□

With Remark 3.38 we notice $fdim(\mathbb{K}) = 2$ and with Theorem 3.39 and Theorem 3.20 we conclude that $\mathfrak{V}$ is planar. □

In [DSRW89, DS05], a modification of the consecutive-one property is introduced that characterizes adjacency matrices of bipartite graphs having a Ferrers-Dimension of at most two:

Theorem 4.21 *[DSRW89] A bipartite graph G has a Ferrers-dimension of at most two if the rows and columns of the biadjacency matrix B (i.e. the cross table) can be permuted independently so that in the rearranged matrix no 0 has a 1 both below it and to its right.*

Since it is shown in [DS05] that this property is equivalent to the Ferrers-graph $\Gamma(B)$ being bipartite we may notice (with Theorem 3.41) that one can characterize planar lattices by this property, too.

However, we will give another characterization that relies more on properties of the corresponding lattice $\mathfrak{V}$, in particular the comparability of certain $\bigwedge$-irreducible elements. By our construction we may even find a left-order of $\mathfrak{V}$.

The consecutive-one property itself of a context $\mathbb{K}$ is not necessary for the planarity of the respective lattice $\mathfrak{V}$. We will show that in a small example. Consider the lattice given in Figure 4.6. Although it is obviously planar, we can not find an enumeration of the $\bigwedge$-irreducibles m_1, m_2, m_3, m_4, s.t. the property is satisfied. This is due to the fact that for each enumeration ε, we find one $\bigwedge$-irreducible m_i, $i \in \{2, 3, 4\}$, s.t. $|\varepsilon(m_1) - \varepsilon(m_i)| > 1$, resulting in a "hole" between m_1 and m_i in the row according to j_i.

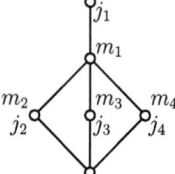

$\mathbb{K}$	m_1	m_2	m_3	m_4
j_1				
j_2	×	×		
j_3	×		×	
j_4	×			×

Figure 4.6: A planar lattice $\mathfrak{V}$ and its standard context $\mathbb{K}$, which does not fulfill the consecutive-one property.

However, we observe that in this particular example the $\bigwedge$-irreducible m_1 is greater than any of the others. Indeed in general we can relax the consecutive-one property by allowing certain "holes". More precisely, any cross $j \leq n$ on the right of a hole $j \nleq m$ must refer to a $\bigwedge$-irreducible n that is greater than m.

This will allow us to characterize planar contexts in a satisfactory way. Besides that an appropriate enumeration already determines a left-order (providing a possibility to actually draw a plane diagram of the lattice according to Section 5.3) as we will point out in the following.

Definition 4.22 *[Zsc06a] Let $\mathbb{K} = (J, M, \leq)$ be the standard context of a lattice $\mathfrak{V}$. Let ε be an enumeration of the attributes satisfying*

$$j \leq m_i, j \nleq m_j, j \leq m_k \quad \Longrightarrow \quad m_j \leq m_k \qquad (4.4)$$

for all $\bigvee$-irreducibles $j \in J$ and all $\bigwedge$-irreducibles $m_i, m_j, m_k \in M$ fulfilling $1 \leq \varepsilon(m_i) < \varepsilon(m_j) < \varepsilon(m_k) \leq |M|$. Then we call ε planar enumeration.

The context $\mathbb{K}$ fulfills the third planarity property (TPC) if it possesses a planar enumeration ε. ◇

Before giving evidence to the fact that the TPC indeed characterizes planarity, we have to clarify how a planar enumeration determines a left-order.

4.4 Planar Contexts

For this purpose we define the relations

$$S_\varepsilon := \{(m,n) \mid m^* = n^* \text{ and } \varepsilon(m) < \varepsilon(n)\}, \tag{4.5}$$

$$L_\varepsilon := \{(m,n) \mid m \parallel n \text{ and } \varepsilon(m) < \varepsilon(n)\}. \tag{4.6}$$

An example of a standard context $\mathbb{K}$ of a lattice $\mathfrak{A}$ possessing a planar enumeration ε is depicted in Figure 4.7. The elements of $M(\mathfrak{A}) = \{n_1, \ldots, n_5\}$ and $J(\mathfrak{A}) = \{j_1, \ldots, j_5\}$ are labeled in the lattice diagram. The enumeration is given by $\varepsilon(n_i) = i$, i.e. by the sorting of the $\bigwedge$-irreducibles in the cross table. In the rows corresponding to the $\bigvee$-irreducibles j_1, j_2, j_4 and j_5 we have consecutive crosses. In the remaining row we find one "hole" (marked). However, $n_3 < n_4$ holds for the involved $\bigwedge$-irreducible elements assuring ε to be a planar enumeration indeed. We notice $S_\varepsilon = \{(n_2, n_4)\}$ and L_ε is given by the table in Figure 4.7.

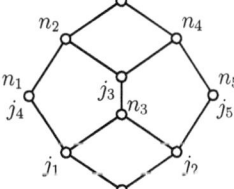

$\mathbb{K}$	n_1	n_2	n_3	n_4	n_5
j_1	×	×	×	×	
j_2		×	×	×	×
j_3			×	□	×
j_4	×	×			
j_5				×	×

L_ε	n_1	n_2	n_3	n_4	n_5
n_1			×	×	×
n_2				×	×
n_3					×
n_4					
n_5					

Figure 4.7: The lattice $\mathfrak{A}$ given by the diagram on the left possesses a standard context $\mathbb{K}$ fulfilling the TPC. The relation L_ε given in the right table determines a left-order L on $\mathfrak{A}$.

Indeed, the relation S_ε provides a sorting relation on $\mathfrak{A}$ and L_ε its left-relation (restricted to $M \times M$) that can be shown to be a left-order with the help of the FPC.

Lemma 4.23 *Let $\mathbb{K} = (J, M, \leq)$ be the standard context of a lattice $\mathfrak{A}$ possessing a planar enumeration ε. Then the following holds:*

1. *S_ε is a sorting relation.*

2. *Let L be the left-relation induced by S_ε, then $S_\varepsilon \subseteq L_\varepsilon = L \cap (M \times M)$.*

3. *L is a left-order.*

Proof:

1. One can easily check that S_ε is both asymmetric and transitive, i.e. a strict order. Additionally, for any two distinct $\bigwedge$-irreducibles $m, n \in M$ sharing the same upper cover m^* we note either $\varepsilon(m) < \varepsilon(n)$ or $\varepsilon(n) < \varepsilon(m)$, i.e. $m\ S_\varepsilon\ n$ or $n\ S_\varepsilon\ m$. According to Definition 4.1 S_ε is a sorting relation.

2. We first note $m^* = n^* \implies m \parallel n$, i.e. $S_\varepsilon \subseteq L_\varepsilon$.

   Now let $m_1, m_2, m_3 \in M$ with $\varepsilon(m_1) < \varepsilon(m_2) < \varepsilon(m_3)$ and $m_1 < m_3 \not> m_2$. We find

   $$\forall j \in J : j \leq m_1 \stackrel{m_1 < m_3}{\implies} j \leq m_3 \stackrel{\text{Def. of }\varepsilon}{\implies} j \leq m_2,$$

   i.e. $m_1 < m_2$. In the same manner one can show that the implication $m_1 > m_3 \not> m_2 \implies m_3 < m_2$ holds for all such $\bigwedge$-irreducibles.

   Let m, n be $\bigwedge$-irreducibles satisfying $m\ L_\varepsilon\ n$ and $(\tilde{m}, \tilde{n}) \in M(m, n)$ be arbitrarily chosen (see picture on the right). We note $\varepsilon(m) < \varepsilon(n)$. The inequality $\varepsilon(\tilde{n}) < \varepsilon(m) < \varepsilon(n)$ implies $n < m$ (see last paragraph) which contradicts $n \parallel m$. We conclude $\varepsilon(m) < \varepsilon(\tilde{n})$. Analogously, $\varepsilon(m) < \varepsilon(\tilde{n}) < \varepsilon(\tilde{m})$ implies $m < \tilde{n}$ contradicting $m \parallel \tilde{n}$. That is, $\varepsilon(\tilde{m}) < \varepsilon(\tilde{n})$ and hence $\tilde{m}\ L_\varepsilon\ \tilde{n}$.

   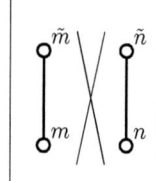

   Now let $(m, n) \in L \triangle L_\varepsilon$ be maximal³. Recall that $m \parallel n$. Let w.l.o.g. $(m, n) \in L$ and $(n, m) \in L_\varepsilon$. According to Definition 4.2 there exists a pair $(\tilde{m}, \tilde{n}) \in M(m, n)$ satisfying $\tilde{m}\ L\ \tilde{n}$. However, due to the last paragraph we conclude $\tilde{n}\ L_\varepsilon\ \tilde{m}$ contradicting the maximality of (m, n). Hence $L \triangle L_\varepsilon = \emptyset$, i.e. $L_\varepsilon = L \cap (M \times M)$.

3. Let $m_1, m_2, m_3 \in M$ satisfying $m_1\ L\ m_2\ L\ m_3$. With the result of the last claim we conclude $m_1\ L_\varepsilon\ m_2\ L_\varepsilon\ m_3$. By applying Definition 4.22 we notice $m_2 > (m_1 \wedge m_3)$. This means that L fulfills the FPC (see Definition 4.13 and Lemma 4.4) and therefore is a left-order (see Proposition 4.14 and Theorem 4.9).

□

Lemma 4.23 gives a possibility to construct a left-order out of a planar enumeration. We still have to show that the TPC is also necessary for planarity of the respective lattice. The following result has been partially published in [Zsc06a]. However, in this version we explicitly emphasize the role of ε for a respective left-order.

---

³With △ we denote the usual symmetric difference.

4.4 Planar Contexts

Theorem 4.24 *Let $\mathbb{K}$ be the standard context of a lattice $\mathfrak{V}$. Then the following holds:*

1. *$\mathfrak{V}$ is planar if and only if $\mathbb{K}$ fulfills the TPC.*

2. *Moreover, every planar enumeration ε of $\mathbb{K}$ uniquely defines a left-order L on $\mathfrak{V}$.*

Proof:

1. $\Leftarrow$: Follows from Lemma 4.23.

   $\Rightarrow$: Let $\mathfrak{V} = (\mathfrak{V}, \leq)$ be planar. Then we find a left-order L on $\mathfrak{V}$. The relation $\tilde{L} := (L \cup \leq) \cap (M \times M)$ is a linear order on M (see Lemma 3.18). Let $\varepsilon : M \to \{1, \ldots, |M|\}$ be an order isomorphism between $(M, \tilde{L})$ and $\{1, \ldots, |M|\}, \leq_\mathbb{N}\}$ (where $\leq_\mathbb{N}$ denotes the standard order on $\mathbb{N}$). Obviously ε exists since both structures are $|M|$-elemental linear ordered sets. By construction, ε is an enumeration of the attributes. Let $j \in J$ and $m_i, m_j, m_k \in M$ be arbitrary with $m_i \,\tilde{L}\, m_j \,\tilde{L}\, m_k$, i.e. $1 \leq \varepsilon(m_i) < \varepsilon(m_j) < \varepsilon(m_k) \leq |M|$. Then we conclude as follows:

$$\begin{aligned} j \leq m_i, j \not\leq m_j, j \leq m_k &\implies m_j \not\geq (m_i \wedge m_k) \\ &\stackrel{FPC}{\implies} m_i \not\!\!L\, m_j \text{ or } m_j \not\!\!L\, m_k \\ &\stackrel{\text{Def. } \tilde{L}}{\implies} m_i <. m_j \text{ or } m_j <. m_k \\ &\stackrel{m_i \not\leq m_j}{\implies} m_j < m_k. \end{aligned}$$

   Therefore ε is a planar enumeration.

2. Every enumeration ε on $\mathbb{K}$ defines uniquely a sorting relation S_ε (see (4.5)) and Lemma 4.23). Furthermore, S_ε uniquely defines a left-relation L (see Definition 4.2) that is a left-order (see Lemma 4.23). $\square$

Finally we want to visualize the connections between the different concepts diagrammatically (see picture on the right). The operators r_1, r_2 and r_3 are restrictions given by

$$\begin{aligned} r_1 &: L \mapsto L \cap (M \times M), \\ r_2 &: L_\varepsilon \mapsto L_\varepsilon \cap M_s, \\ r_3 &: L \mapsto L \cap M_s, \end{aligned}$$

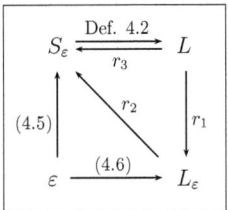

where $M_s := \{(m,n) \mid m^* = n^*\}$ denotes the set of pairs of $\bigwedge$-irreducibles sharing the same upper neighbor.

Note that a left-order L on a lattice $\mathfrak{V}$ does not determine a unique planar enumeration ε. In particular, L does not order comparable elements. An n-elemental chain C_n for instance possesses exactly one left-order $L = \emptyset$ but $(n-1)!$ planar enumerations since any bijective mapping ε of the $n-1$ $\bigwedge$-irreducibles into the set $\{1, \ldots, n-1\}$ fulfills condition (4.4).

Unfortunately the TPC does not supply an efficient way of finding one or all plane diagrams of a lattice nor stating that it is not planar. Although there exists a linear-time algorithm in terms of the order relation $\leq_\mathfrak{V}$ of the lattice (i.e. time complexity is $\mathcal{O}(|\leq_\mathfrak{V}|) \leq \mathcal{O}(|J(\mathfrak{V})| \cdot |M(\mathfrak{V})|)$) to find a consecutive-one arrangement of a context if it exists [BL76], for the more general TPC we do not know about such an algorithm.

However, for a small example one can often find a planar enumeration by hand. See Figure 4.8 for a flavor of how it can be detected.

To get a starting configuration, it is always a good idea to put the greater $\bigwedge$-irreducibles to the right and the smaller to the left in a respective cross table. By pushing the coatoms n_1 and n_2 to the right, n_3 and n_4 into the middle and the remaining $\bigwedge$-irreducibles n_5, n_6 and n_7 to the left, we have nearly found ε (left table). The two problematic holes in the last row (since n_1 and n_3 are not lesser than n_2) can be closed if one shifts n_4 and n_7 right of n_1 and n_3 (table in the middle). The remaining obstacles (marked) can be removed by exchanging the columns corresponding to n_6 and n_3. The holes in the emerging cross table (on the right) do not break condition 4.4 since we have $n_6 < n_1, n_2$ and $n_7 < n_4 < n_2$. The resulting planar enumeration ε supplies the sorting relation $S_\varepsilon = \{(n_1, n_2)\}$ and consequently a left-order L by applying Definition 4.2.

$\mathbb{K}$	n_5	n_6	n_7	n_3	n_4	n_1	n_2
j_1	×			×	×		
j_2			×		×	×	
j_3		×			×	×	
j_4				×	×	×	
j_5		×	□	×	□		×

$\mathbb{K}$	n_5	n_6	n_3	n_1	n_7	n_4	n_2
j_1	×	□	×	×			
j_2		×	×				×
j_3	×	□	×				×
j_4					×	×	×
j_5					×	×	×

$\mathbb{K}$	n_5	n_3	n_6	n_1	n_7	n_4	n_2
j_1	×	×		×			
j_2		×		×			×
j_3			×	×			×
j_4				×		×	×
j_5					×	×	×

Figure 4.8: Constructing a planar enumeration of a context $\mathbb{K}$.

Determining a planar enumeration out of a planar lattice $\mathfrak{V}$ is much simpler however, if one knows a left-order L on $\mathfrak{V}$ (for instance, it can be read from a diagram, see Figure 5.5). By restricting L on pairs of $\bigwedge$-irreducibles and enriching it to a linear extension, a planar enumeration is found. See Figure

4.4 Planar Contexts

4.9 for an example. On the left, the lattice according to the context of Figure 4.8 is given by a plane diagram. The $\bigwedge$-irreducibles and $\bigvee$-irreducibles are labeled. We gain a linear extension by adding the strict order relation $<$ of the lattice itself. Sorting the $\bigwedge$-irreducibles according to this linear order supplies a planar enumeration η. Indeed, in this case η and ε are equal.

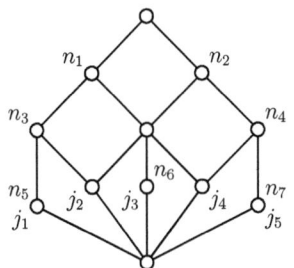

$\tilde{L}$	n_5	n_3	n_6	n_1	n_7	n_4	n_2
n_5		$<$	L	$<$	L	L	L
n_3			L	$<$	L	L	L
n_6				$<$	L	L	$<$
n_1					L	L	L
n_7						$<$	$<$
n_4							$<$
n_2							
$\eta(x)$	1	2	3	4	5	6	7

Figure 4.9: Constructing a planar enumeration out of a left-order on a lattice $\mathfrak{V}$. The relation $\tilde{L} := (L \cup \leq) \cap (M \times M)$ was used already in the proof of Theorem 4.24.

Chapter 5

Left-relations on Diagrams

In the last chapter we introduced left-relations on lattices. This is remarkable since the notion of "left" suggests a visual aspect of the relation which is, however, associated with a rather abstract construction. Additionally, we based some "intuitive" explanations on diagrams where the concepts left and right are somehow evident but not formally defined.

In this chapter we want to close these gaps. Similarly to the previous one, we introduce sorting relations and left-relations that are defined on (or rather induced by) *diagrams* however. Thereby we again base our method on ideas given in the work of Kelly and Rival [KR75].

While the definition of the sorting relation remains effectively the same, left-relations are introduced in a different way. An inductive approach as in Definition 4.2 would appear clumsy; in a given diagram one expects to be able to "see" directly which elements are left or right of others.

However, at least in case of planarity, there exists a strong coherence between both concepts. It will be explained in Section 5.2. In fact every left-order determines a plane diagram and, vice versa, a plane diagram represents a conjugate order.

This enables us to actually create plane diagrams of planar lattices in Section 5.3. We will even specify the construction principle of the *left-right-numbering* to draw attribute additive plane diagrams out of a left order and we will argue why this algorithm may fail for layer diagrams.

5.1 Definition

In Definition 4.1 the set of $\bigwedge$-irreducibles being lower neighbors of a certain lattice element are ordered from left to right. This is actually motivated by a view on a respective drawing, where such an order can easily be recognized, namely by the angle of the line connecting a node with the upper neighbor. An example is given in Figure 5.1.

Definition 5.1 *[Zsc05]* Let $\mathrm{pos}(\mathfrak{V})$ be diagram of the lattice $\mathfrak{V}$. For each $\bigwedge$-irreducible $n \in M$, let $\varphi(\mathrm{pos}(n\,n^*))$ denote the angle between a horizontal line through n^* and the line $\mathrm{pos}(n\,n^*)$. The binary relation $\sigma \subseteq M \times M$ defined by
$$m \; \sigma \; n :\iff m^* = n^* \text{ and } \varphi(\mathrm{pos}(m\,m^*)) < \varphi(\mathrm{pos}(n\,n^*)).$$
is called sorting relation induced by $\mathrm{pos}(\mathfrak{V})$. $\diamond$

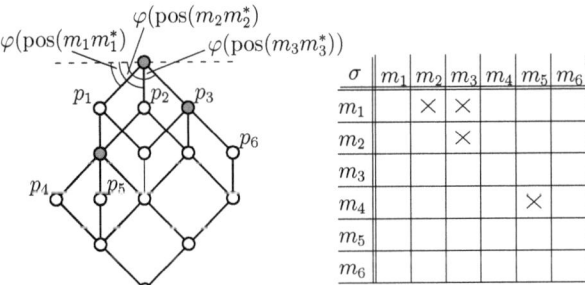

Figure 5.1: A diagram of a lattice with its sorting relation. The symbols $p_i := \mathrm{pos}(m_i)$ ($i = \{1, \ldots, 6\}$) label the nodes of the $\bigwedge$-irreducibles m_i. The gray points represent the set of upper neighbors of $\bigwedge$-irreducible elements.

In the following we will, as already done beforehand, identify lattice elements v and their diagram nodes $\mathrm{pos}(v)$ for conveniency.

Obviously every sorting relation σ on a diagram is a sorting relation on the underlying lattice since the angles of the respective diagram lines can be ordered strictly. Note that two distinct lines $\mathrm{pos}(v_1 w)$ and $\mathrm{pos}(v_2 w)$ do not have the same angle since this contradicts condition 4 of Definition 3.10. Conversely it is easy to realize a sorting relation S in a diagram.

The last definition reminds us of the structure of the left-relation given by Kelly and Rival (see Definition 3.31). There are two differences. We consider also non-plane diagrams but restrict the sorting relation (which can be seen as λ^* in their approach) to $\bigwedge$-irreducible elements.

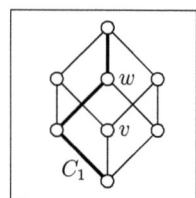 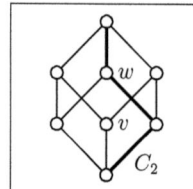

Figure 5.2: In the left picture we observe $v \varrho w$ since v is in the area right of the maximal chain C_1 (marked). In analogy, we find $v \lambda w$ (see right picture).

Now we are going to to introduce left-relations on diagrams (see Figure 5.2). In contrast to the last chapter, they are not induced by a sorting relation (although they contain exactly one) since even fixing the vectors of the $\bigwedge$-irreducibles restricts us only little in the layout of the remaining of the drawing.

Definition 5.2 *[KR75, Zsc05] Let $\mathfrak{V}$ be a finite lattice and $\mathrm{pos}(\mathfrak{V})$ a diagram of it. For a maximal chain C,*

$$F_l(C) := \{(x,y) \in \mathbb{R}^2 \mid y \in [y(0_\mathfrak{V}), y(1_\mathfrak{V})], x < x_C(y)\}$$

is the area left of $\mathrm{pos}(C)$ and dually $F_r(C)$ the area right of $\mathrm{pos}(C)$. We define the left- and the right-relation λ and ϱ induced by $\mathrm{pos}(\mathfrak{V})$ by

$$v \lambda w \quad :\iff \quad (\exists C \ni w : \mathrm{pos}(v) \in F_l(C)) \wedge (v \parallel w),$$
$$v \varrho w \quad :\iff \quad (\exists C \ni w : \mathrm{pos}(v) \in F_r(C)) \wedge (v \parallel w)$$

for all elements $v, w \in \mathfrak{V}$. ◇

In the following remark we compile some basic properties of the relations λ and ϱ and their defining sets $F_l(C)$ and $F_r(C)$.

Remark 5.3 *Let $\mathfrak{V}$ be a lattice with a diagram $\mathrm{pos}(\mathfrak{V})$. Let λ and ϱ be the left- and right-relation on $\mathrm{pos}(\mathfrak{V})$. Let C be a maximal chain in $\mathrm{pos}(\mathfrak{V})$. Then the following hold:*

1. $F_l(C) \;\dot{\cup}\; \mathrm{pos}(C) \;\dot{\cup}\; F_r(C) = F(\mathrm{pos}(\mathfrak{V})) := \mathbb{R} \times [y(0_\mathfrak{V}), y(1_\mathfrak{V})]$.

2. $\lambda \cup \varrho = \parallel$.

3. $\lambda \cup \lambda^{-1} \subseteq \parallel$.

4. *If $\mathrm{pos}(\mathfrak{V})$ is a plane line diagram then σ is the restriction of the left-relation λ to pairs of $\bigwedge$-irreducibles with common upper neighbor, i.e.*

$$m^* = n^* \text{ and } m \lambda n \iff m \sigma n.$$

5.1 Definition

5. There exist diagrams sharing the same sorting relation σ but possessing different left-relations.

Proof:

1. This is due to the fact that the corresponding function $x(C)$ is indeed a function, i.e. its domain is $[y(0_\mathfrak{V}), y(1_\mathfrak{V})]$ and $x(y)$ is a unique point.

2. Let v and w be arbitrary incomparable elements of $\mathfrak{V}$. There exists a maximal chain $C = 0_\mathfrak{V} \ldots w \ldots 1_\mathfrak{V}$. Clearly $\text{pos}(v) \in F(\text{pos}(\mathfrak{V})) \setminus \text{pos}(C)$, the assertion follows with 1. and Definition 5.2.

3. Except for some degenerated diagrams left and right side of the inequality are equal. See the picture on the right for an example. There exists exactly one maximal chain containing v and w is in the area right of this chain, i.e. $v \varrho w$. In the same way we conclude $w \varrho v$.

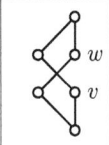

4. Let m and n be $\bigwedge$-irreducibles with common upper neighbor m^*. Let x_{mm^*} and x_C be the corresponding functions of the line $\text{pos}(mm^*)$ and the image of a maximal chain C containing n. Since these lines do not intersect, we conclude with the intermediate value theorem either $x_C(y) < x_{mm^*}(y)$ or $x_C(y) > x_{mm^*}(y)$ for all $y \in [y(m), y(m^*)]$. Hence $m \sigma n \iff m \lambda n$.

5. An example for this fact is given in Figure 5.3. □

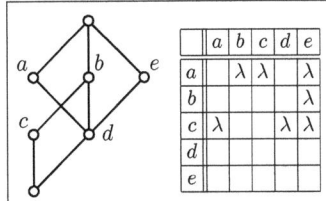 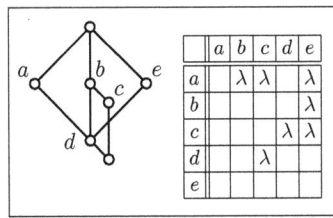

Figure 5.3: Two diagrams of the same lattice with their left-relations. Both include the sorting relation $a \sigma b \sigma e$ However, their left-relations are not equal, as depicted in the respective tables.

5.2 Coherence to Left-relations on Lattices

As already mentioned, the introduced concepts resemble appropriate approaches to add a left-right perspective to the standard up-down one in diagrams given by Kelly and Rival [KR75] in Definition 3.31. Now we want to merge both frameworks of left-relations. This will finally enable us to compare left-relations on lattices and on diagrams.

We start with a lemma that confirms the containedness relations between σ, λ and λ^* one may expect already.

Lemma 5.4 *Let* pos($\mathfrak{V}$) *be a plane diagram of the lattice* $\mathfrak{V}$. *Let* σ *be its sorting relation, λ its left-relation and λ^* the relation according to Definition 3.31. Then $\sigma \subseteq \lambda^* \subseteq \lambda$.*

Proof: $\sigma \subseteq \lambda^*$ is an immediate consequence of the Definitions 3.31 and 5.1.

$\lambda^* \subseteq \lambda$: We repeat the proof originally published in [KR75]: Let $v \, \lambda^* \, w$ and

$$C = 0 \prec w_1 \prec \ldots \prec w_r \prec w \prec w^*$$

be a chain. Since pos($\mathfrak{V}$) is a diagram, we know by Definition 3.31 that $x_1 := x_{vw^*}(y(w)) < x(w)$. By iteratively applying Lemma 3.30 we conclude $x_{vw^*}(y(w_k)) < x(w_k)$ for all k with $y(w_k) > y(v)$. This implies $x(v) < x_C(y(v))) =: x_2$, i.e. $v \, \lambda \, w$ (see picture on the right).

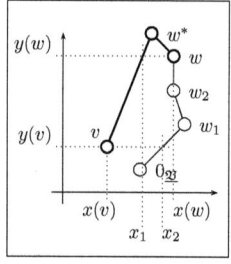

Now we are able to bring together the ideas developed in Definitions 3.31, 5.1 and 5.2, at least for plane diagrams pos($\mathfrak{V}$) of lattices $\mathfrak{V}$. The first assertion of Corollary 5.5 claims that in that case the relation "to the left of" and the left-relation are equal. Consequently one can define similarity in terms of appropriate left-relations and discover them to be left-orders.

While these three conclusions were more or less known (see [KR75]), the remaining two involve the much smaller sorting relation σ into our investigation. Both σ and λ uniquely determine each other in an analog way as sorting and left-relations do in the lattice case.

Remarkably is the last statement. It characterizes similarity of diagrams just by the equality of the underlying sorting relations instead of the larger relation λ^*. This means that the example depicted in Figure 5.3 can not occur. Two plane diagrams possessing the same sorting relation have the same left-relation, too.

5.2 Coherence to Left-relations on Lattices

Corollary 5.5 *Let $\mathfrak{V}$ be a planar lattice. Then the following holds:*

1. *[KR75] The left-relation λ of a plane diagram $\mathrm{pos}(\mathfrak{V})$ is equal to its "to the left" relation (see Definition 3.31).*

2. *Two (plane) diagrams are similar if and only if their respective left-relations λ are equal.*

3. *The left-relation λ of a plane diagram $\mathrm{pos}(\mathfrak{V})$ is a left-order on $\mathfrak{V}$.*

4. *In a plane diagram $\mathrm{pos}(\mathfrak{V})$, the left-relation λ and the sorting relation σ uniquely determine each other. Moreover, λ can be calculated from σ via the construction given in Definition 4.2.*

5. *Two (plane) diagrams are similar if and only if their respective sorting relations σ are equal.*

Proof:

2. By definition, two plane diagrams $\mathrm{pos}_1(\mathfrak{V})$ and $\mathrm{pos}_2(\mathfrak{V})$ are similar if their respective relations λ_1^* and λ_2^* correspond to each other. Hence the relations "to the left of" are equal. With statement 1 we conclude that also the left-relations are equal.

   On the other hand, if two diagrams possess equal left-relations $\lambda_1 = \lambda_2$ then $\lambda_1^* = \lambda_1^*$ holds for the respective relations according to Lemma 5.4. Hence the diagrams are similar.

3. This follows from Corollary 5.5 1., Proposition 3.32 and Theorem 4.9.

4. Let $\mathrm{pos}(\mathfrak{V})$ be a plane diagram of $\mathfrak{V}$ with sorting relation σ and left-relation λ. Due to 3. λ is a left-order, i.e. uniquely determined (see Remark 4.5 2.) by a sorting relation $S = \lambda \cap \{(m,n) \mid m^* = n^*\}$. Clearly, $\sigma = S$ since $\sigma \subseteq \lambda \cap \{(m,n) \mid m^* = n^*\}$.

5. Two similar diagrams have the same relation λ^* and hence (see Lemma 5.4) the same sorting relation.

   Conversely, two diagrams possessing the same sorting relation $\sigma_1 = \sigma_2$ also possess the same left-relation $\lambda_1 = \lambda_2$ (see 4.), i.e. are similar (see 2.). □

The following theorem connects left-relations on lattices and those on diagrams: Every left-relation on a plane diagram $\mathrm{pos}(\mathfrak{V})$ is a left-order on $\mathfrak{V}$, and vice versa every left-order L on $\mathfrak{V}$ can be realized in a plane diagram. Consequently we can characterize all plane diagrams (up to similarity) by a lattice theoretic condition without taking care of topological properties. Keep in mind that parts of the following result are to be found in Theorem 3.20 already.

Theorem 5.6 *[Zsc05] Let $\mathfrak{V}$ be a finite lattice. The following statements are equivalent.*

1. *There exists a plane diagram $\text{pos}(\mathfrak{V})$ with the left-relation L.*

2. *L is a left-order on $\mathfrak{V}$.*

Proof:
"1. $\Rightarrow$ 2.": That is the assertion of Corollary 5.5 3.
"2. $\Rightarrow$ 1.": We define two relations $L_< := L \cup <$ and $R_< := R \cup <$. It is easy to show (see Theorem 3.20) that they are linear orders. Let the maps l and r be embeddings of $(\mathfrak{V}, L_<)$ and $(\mathfrak{V}, R_<)$ respectively into $(\mathbb{R}, <)$.

Let pos be a map assigning to each $v \in \mathfrak{V}$ the point $(l(v), r(v))$ and to each pair of neighboring elements a straight line segment connecting them. By the definitions of l and r we realize that pos meets the conditions 1, 2 and 3 of Definition 3.10. We show now that no line segments cross which makes the image of pos a plane line diagram of $\mathfrak{V}$.

We assume that the diagram edges corresponding to the elements $v_1 \prec v_3$ and $v_2 \prec v_4$ cross. Let (x_i, y_i) be the coordinates of the node v_i and (x_5, y_5) be the coordinates of the intersection. Since l and r are order homomorphisms we observe $x_1, x_2 < x_5 < x_3, x_4$ and $y_1, y_2 < y_5 < y_3, y_4$. We conclude $v_2 < v_3$ and $v_1 < v_4$ and therefore $v_1 \parallel v_2$ and $v_3 \parallel v_4$. That means that v_3 and v_4 do not have an infimum in contradiction to $\mathfrak{V}$ being a lattice.

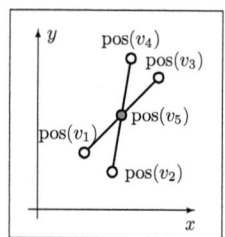

Let λ be the induced left-relation by $\text{pos}(\mathfrak{V})$. We finally show that $\lambda = L$ holds. Let $m\ S\ n$ hold for some $m, n \in M$. Due to the definitions of r and l the inequalities $x(m) < x(n) < x(m^*)$ and $y(n) < y(m) < y(m^*)$ hold. For the angles $\varphi_m := \varphi(\text{pos}(m\,m^*))$ and $\varphi_n := \varphi(\text{pos}(n\,n^*))$ we get

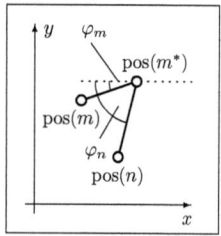

$$\tan \varphi_m = \frac{y(m) - y(m^*)}{x(m) - x(m^*)}, \quad \tan \varphi_n = \frac{y(n) - y(m^*)}{x(n) - x(m^*)}.$$

Since the inequalities

$$y(n) - y(m^*) < y(m) - y(m^*) < 0 \text{ and } x(m) - x(m^*) < x(n) - x(m^*) < 0$$

hold, we conclude $\tan \varphi_m < \tan \varphi_n$. Both angles are in the interval $(0, \pi/2)$. In this domain the function arctan is monotonous, so we conclude $\varphi_m < \varphi_n$, i.e.

5.2 Coherence to Left-relations on Lattices

$m \ \sigma \ n$. This means that the sorting relations of the lattice and its diagram are the same.

Due to Corollary 5.5, 4., we can conclude $\sigma = S \implies \lambda = L$ since λ is the left-relation of a plane diagram. □

Corollary 5.7 *Let $\mathfrak{V}$ be a lattice. Then the number of non-similar plane diagrams is equal to the number of left-orders on $\mathfrak{V}$.*

Proof: This follows immediately from Theorem 5.6 and Corollary 5.5 2. □

It is an interesting question whether the result of Theorem 5.6 could be extended: instead of left-*orders* and plane diagrams we want to investigate interconnections between left-*relations* on diagrams and lattices. We already know that some left-relations on diagrams can not be derived as left-relations of lattices. This is a consequence of Remark 5.3 3. The other direction remains unsolved:

Conjecture: For every left-relation L on $\mathfrak{V}$ there exists a diagram $\text{pos}(\mathfrak{V})$ possessing the left-relation L.

We want to recapitulate the coherences between sorting and left-relations visually in Figure 5.4. Two symbols ./. marking an arrow refer to different settings for the planar and the general case. A left relation L is uniquely defined by its sorting relation S (see Remark 4.5), vice versa S can be found by restricting L to pairs of $\bigwedge$-irreducibles sharing the same upper cover. The restricting process works as well for the diagram relations λ and σ (Remark 5.3 4.).

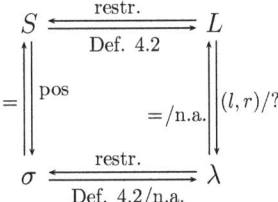

Figure 5.4: Connections between left- and sorting relations on lattices and diagrams.

However, Definition 4.2 can be used to build λ out of σ only in case of plane diagrams (see Corollary 5.5 4.). In general, λ is not uniquely determined by σ (see Remark 5.3 5. and Figure 5.2).

Let us now consider the arrows on the left: every sorting relation σ can indeed be understood as a sorting relation S according to Definition 4.1 since it is a collection of linear orders on sets of $\bigwedge$-irreducibles with common upper neighbor. Dually, any sorting relation S can be realized in a diagram; just draw all $\bigwedge$-irreducibles according to their order from left to right (keeping in mind the upward drawing constraint Definition 3.10, 2.) and add the remaining diagram nodes and lines.

Eventually, things on the right side are most complicated: a left-relation λ of a plane diagram is a left-order (see Corollary 5.5, 3.). In general however, λ does not define a left-relation L (see Remark 5.3, 3.). From a left-order L one can construct a diagram possessing the left-relation L by the mapping pos : $v \mapsto (l(v), r(v))$ according to the proof of Theorem 5.6. In general it is unclear whether an arbitrary left-relation L can be found in a diagram (see the conjecture in the last paragraph).

Finally let us explain in an example, how to read the left-relation λ of a plane diagram. This can be done in the non-plane case too, however, we are interested in the emerging left-order $L = \lambda$. In general it may be easier (and more useful) to obtain the larger linear order $L_< = L \cup <$ and then to remove pairs of comparable elements. Consider Figure 5.5 for an example.

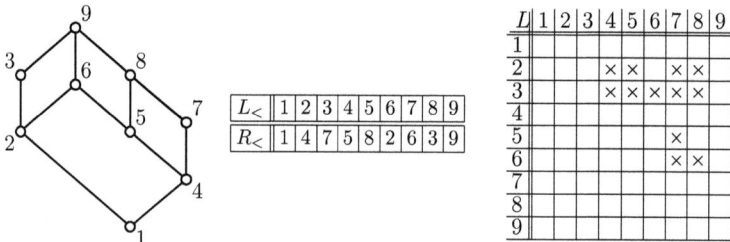

Figure 5.5: A plane diagram of a lattice together with the linear orders $L_<$ and $R_<$ and the left-order L.

The order $L_<$ roughly runs from the lower left to the upper right. The lowest element is the bottom element 1 of the lattice. Next come elements 2 and 3 on the left edge. Imagine to delete the first three elements of the diagram (seen as a graph diagram). Then 4 is the smallest element and comes next in the hierarchy therefore. We continue with the smaller 5 and 6 until the elements 7, 8 and 9 on the upper right edge complete the linear order (see table in the middle). In an analog way one can calculate $R_<$. Deleting comparable pairs finally gives the left-order $L = R^{-1}$ depicted by a relation table on the right.

5.3 How to Draw Plane Diagrams

In the last section we gave in the proof of Theorem 5.6 a method to draw plane diagrams of planar lattices $\mathfrak{V}$ if we know a left-order L on $\mathfrak{V}$ (for some examples, see e.g. [Tro92]). We will call it, following the notation in [CT94], *left-right-numbering*. It is done in three steps:

1. Extend L to two linear orders (realizers due to Definition 3.5)

$$L_< := L \cup < \quad \text{and} \quad R_< := R \cup < .$$

2. Provide two embeddings (which we will call *diagram realizers*)

$$l : (\mathfrak{V}, L_<) \hookrightarrow (\mathbb{R}, <) \quad \text{and} \quad r : (\mathfrak{V}, R_<) \hookrightarrow (\mathbb{R}, <).$$

3. Apply the point $\text{pos}(v) := (l(v), r(v))$ to each lattice element v and connect adjacent nodes.

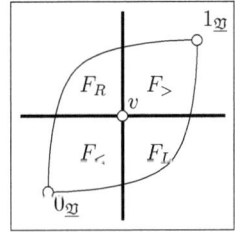

In a diagram arising from this construction, the top element of the lattice is mapped to the upper right corner and the bottom element to the lower left. Let $v \in \mathfrak{V}$ then the elements left, right, above and below v can easily be encountered in the drawing. The lines

$$x_v := \text{pos}(v) + \{(r, 0)^T \mid r \in \mathbb{R}\} \quad \text{and}$$
$$y_v := \text{pos}(v) + \{(0, r)^T \mid r \in \mathbb{R}\}$$

divide the plane $\mathbb{R}^2$ into the four areas $F_>$, F_R, F_L and $F_<$ (see picture). Due to the properties of the mappings l and r, we notice

$$\begin{aligned} \text{pos}(w) \in F_> &\iff w > v, & \text{pos}(w) \in F_< &\iff w < v, \\ \text{pos}(w) \in F_R &\iff w \, R \, v, & \text{pos}(w) \in F_L &\iff w \, L \, v. \end{aligned} \quad (5.1)$$

Diagrams of posets fulfilling property (5.1) are called *dominance drawings* [BCB+92, CT94]. Every poset of order dimension n can be embedded as a dominance drawing into an n-dimensional grid [EEH+97], this is a consequence of the equivalence of order and product dimension (see Definition 3.6 and Theorem 3.20). For an example of a dominance drawing have a look at Figure 3.4

When we layout a diagram we want to implement additional constraints, for instance attribute additivity or the layer diagram convention. In this section, we want to investigate whether we can apply the left-right-numbering-algorithm to this particular methods and specify the maps l and r precisely, if possible.

5.3.1 Attribute Additive Diagrams

In the following we will give a recursive definition of two mappings l and r, s.t. $\text{pos}(v) := (l(v), r(v))$ induces a plane attribute additive diagram of a planar lattice $\mathfrak{V}$ with left-order L. We set

- for the top element $1_{\mathfrak{V}}$ of the lattice

$$l(1_{\mathfrak{V}}) = r(1_{\mathfrak{V}}) = 0, \qquad (5.2)$$

- for each $\bigwedge$-irreducible m we choose positive reals $d_l(m)$ and $d_r(m)$ and set

$$l(m) = l(v) - d_l(m) \quad :\Longleftrightarrow\quad m \prec_{L_<} v \qquad (5.3)$$
$$r(m) = r(v) - d_r(m) \quad :\Longleftrightarrow\quad m \prec_{R_<} v, \qquad (5.4)$$

where $\prec_{L_<}$ and $\prec_{R_<}$ are the neighbor relations of $L_<$ and $R_<$, and

$$\text{vec}_l(m) = l(m) - l(m^*), \quad \text{vec}_r(m) = r(m) - r(m^*) \qquad (5.5)$$

- For any other lattice element v

$$l(v) := \sum_{m \in M(v)} \text{vec}_l(m), \quad r(v) := \sum_{m \in M(v)} \text{vec}_r(m). \qquad (5.6)$$

In Figure 5.6 we give an example for this construction. In order to see that this choice will meet our expectations, we effectively have to evidence that the attribute additive part of the construction (i.e. Equation 5.6 does not destruct the *diagram realizer* property of l and r respectively. We want to mention that unlike in [Zsc05] the diagram order increases to the upper right (instead of upper left) in order to gain a *dominance drawing*.

Theorem 5.8 *[Zsc05] Every finite planar lattice possesses a plane attribute additive diagram.*

Furthermore, if L is a left-order on a finite lattice $\mathfrak{V}$ then there exists a plane attribute additive diagram $\text{pos}(\mathfrak{V})$ with $\text{pos}(v) := (l(v), r(v))$ (as defined in the Equations 5.2-5.6) possessing the left-relation L.

Proof: We have to evidence the following claims:

1. The mapping pos determines a point to each lattice element v: Let $v \neq 1_{\mathfrak{V}}$ be a lattice element, s.t. $l(w)$ is determined for all $v \leq_{L_<} w$. If v is a $\bigwedge$-irreducible, then we find $l(v) = l(\tilde{v}) - 1$ (Equation (5.3)), where $\tilde{v}$ is the upper neighbor of v w.r.t. $L_<$. Otherwise v is not a $\bigwedge$-irreducible element, then $l(m)$ is assigned to all $m \in M(v)$ since $v L_< m$ holds for all $m \in M(v)$. Therefore we can calculate $l(v)$ according to Equation 5.6. A similar argumentation can be applied for the mapping r.

5.3 How to Draw Plane Diagrams

2. The resulting diagram is attribute additive. This can be easily noticed in Equation 5.6 of the definition of l and r.

3. The maps l and r are diagram realizers. Since $L_<$ and $R_<$ are linear orders, we have to show only

$$v \prec_{L_<} \tilde{v} \implies l(v) < l(\tilde{v}) \text{ and } v \prec_{R_<} \tilde{v} \implies r(v) < r(\tilde{v}). \quad (5.7)$$

Let $v \in M$. Then 5.7 follows directly from Equations (5.3) and (5.4). Let otherwise $v \in \mathfrak{V}$ be a non-$\bigwedge$-irreducible element with the upper neighbor (w.r.t. $L_<$) $\tilde{v}$. So either $v \ L \ \tilde{v}$ or $v < \tilde{v}$ holds.

Assume $v \ L \ \tilde{v}$. Since v has at least two upper neighbors $w_1 \ L \ w_2$ we know

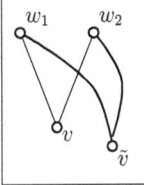

$$v \prec_{L_<} \tilde{v} <_{L_<} w_1 <_{L_<} w_2.$$

Since L is transitive we conclude $\tilde{v} < w_1$ (otherwise $\tilde{v} \ L \ w_1$ would imply $v \ L \ w_1$) and $\tilde{v} < w_2$ respectively. Therefore w_1 and w_2 are incomparable minimal upper bounds of the set $\{v, \tilde{v}\}$ contradicting $\mathfrak{V}$ to be a lattice.

Therefore the inequality $v < \tilde{v}$ holds. This implies $M(v) \supseteq M(\tilde{v})$ and hence

$$l(v) = \sum_{m \in M(v)} \text{vec}_l(m) < \sum_{m \in M(\tilde{v})} \text{vec}_l(m) = l(\tilde{v})$$

since $\text{vec}_l(m) < 0$ holds for all $\bigwedge$-irreducibles m. Analogously we find $r(v) < r(\tilde{v})$.

Hence the mapping pos induces (by connecting all neighbored pairs of lattice element by straight lines) an attribute additive plane diagram of $\mathfrak{V}$. □

Theorem 5.8 allows to construct diagrams pos($\mathfrak{V}$) with the left-relation L from arbitrary left-orders L on $\mathfrak{V}$. This gives rise to the following:

Corollary 5.9 *Every plane diagram is similar to a plane attribute additive diagram.*

Proof: A plane diagram $\text{pos}_1(\mathfrak{V})$ is determined up to similarity by its left-relation λ. Since λ is a left-order, we can construct a plane attribute additive diagram $\text{pos}_2(\mathfrak{V})$ possessing the same left-relation λ. Therefore $\text{pos}_1(\mathfrak{V})$ and $\text{pos}_2(\mathfrak{V})$ are similar. □

Finally we want to describe the construction of the mappings l and r and the emerging diagram on an example. Consider the interordinal lattice $\mathfrak{B}(\mathbb{I}_4)$ depicted on the right. Note that we used a diagram for the visualization but are interested in the underlying lattice only. We recognize six $\bigwedge$-irreducible elements labeled a, b, c, d, e and f and further five elements 0, 1, x, y and z. The left-relation L of this lattice induced by the sorting relation $S = \{(a,b)\}$ is a left-order.

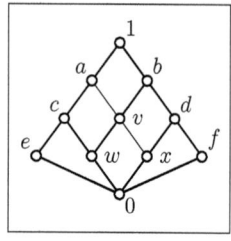

In Figure 5.6 the respective relations $L_<$ and $R_<$ are depicted as well as the mappings l and r, calculated due to the definition from "right to left" (we chose $d_l(m) = d_r(m) = 1$ for all $m \in M$) and the mappings vec_l and vec_r.

$L_<$	0	e	w	c	x	v	a	f	d	b	1
l	-12	-9	-8	-7	-6	-5	-4	-3	-2	-1	0
vec_l		-2		-3			-4	-1	-1	-1	
$R_<$	0	f	x	d	w	v	b	e	c	a	1
r	-12	-9	-8	-7	-6	-5	-4	-3	-2	-1	0
vec_r		-2		-3			-4	-1	-1	-1	

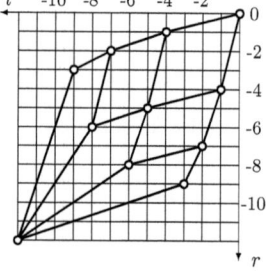

Figure 5.6: Constructing a plane attribute additive diagram of the interordinal lattice $\mathfrak{B}(\mathbb{I}_4)$ via the construction given by the Equations (5.2)-(5.6).

5.3.2 Layer Diagrams

Unfortunately a similar construction as for attribute additive diagrams is not applicable in general for layer diagrams. That means that there exist lattices $\mathfrak{V}$ with a given layer assignment function $p : \mathfrak{V} \to \mathbb{R}$, s.t. no diagram realizers exist. We will show this in the following. For p, we chose the *longest path layering* method [ES90] which is the most common approach.

Let L be a left-order on a planar lattice $\mathfrak{V}$. Let l and r be diagram realizers w.r.t. L and $\mathrm{pos}(\mathfrak{V})$ a diagram satisfying $\mathrm{pos}(v) = (l(v), r(v))$ for all lattice elements v. W.l.o.g. let $l(1) = r(1) = 0$. Obviously the layers can not be modeled as horizontal lines, otherwise two elements v and w represented on the same layer would satisfy $l(v) = l(w)$ contradicting l to be a diagram realizer.

5.3 How to Draw Plane Diagrams

Hence the layers are parallel lines $\alpha \cdot x + \beta \cdot y = c$. Moreover, if two elements $v \, L \, w$ are situated on the same layer, i.e. if

$$\alpha \cdot r(v) + \beta \cdot l(v) = p(v) = p(w) = \alpha \cdot r(w) + \beta \cdot l(w)$$

then we notice $v \, L_< \, w$ and $w \, R_< \, v$, i.e. $l(v) < l(w)$ and $r(w) < r(v)$. We conclude $\alpha \cdot \beta > 0$ and if we assume $c < 0$ then we find $\alpha > 0$ and $\beta > 0$ since $l(v) \leq 0$ and $r(v) \leq 0$ holds for all lattice elements $v \in \mathfrak{V}$.

We consider the interordinal lattice $\mathfrak{B}(\mathbb{I}_4)$ investigated in the last section together with the longest path layer assignment function p given by

v	1	a	b	c	v	d	e	w	x	f	0
$p(v)$	0	-1	-1	-2	-2	-2	-3	-3	-3	-3	-4

and the left-order induced by the sorting relation $S = \{(a,b)\}$. Now we conclude as follows:

$$\alpha \cdot r(a) < 0 \implies \beta \cdot l(a) > -1 \implies \beta \cdot l(f) > -1 \implies \alpha \cdot r(f) < -2$$
$$\beta \cdot l(b) < 0 \implies \alpha \cdot r(b) > -1 \implies \alpha \cdot r(e) > -1 \implies \beta \cdot l(e) < -2,$$

i.e. $\alpha \cdot r(0) + \beta \cdot l(0) < -4 = p(0)$ since we know $r(0) > r(f)$ and $l(0) < l(e)$ according to the left-order L (see picture on the right, the layers are the diagonal lines). This means that the bottom element of the lattice can not be positioned on its desired layer. An analog argumentation can be applied for the second left-order induced by the sorting relation $S = \{(b,a)\}$. We summarize to the following:

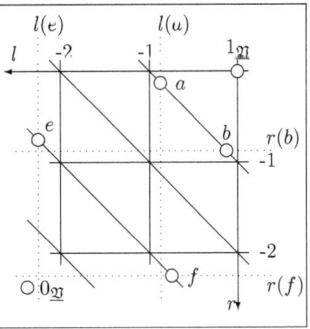

Proposition 5.10 *There exist planar lattices $\mathfrak{V}$ with a longest path layer assignment function $p : \mathfrak{V} \to \mathbb{R}$, s.t. no pair of diagram realizers r and l creating a layer diagram $\text{pos}(\mathfrak{V}) = (r(v), l(v))$ according to p exist.*

Nevertheless we assume that there exists a plane layer diagram for every planar lattice with a layer assignment function p.

5.3.3 Layered Attribute Additive Diagrams

Finally, we want to analyze the case of *layered attribute additive* diagrams (see Definition 3.15). Unfortunately, also this drawing convention will not always allow to find a diagram realizer, as we will evidence in the following.

Consider the 4-elemental dual $\mathfrak{B}(\mathbb{I}_4)^d$ of the interordinal scale $\mathfrak{B}(\mathbb{I}_4)$ we observed already in the last two sections. It contains the $\bigwedge$-irreducibles $m_1, \ldots, m_4$ and further elements $1, v_{12}, v_{13}, v_{14}, v_{23}, v_{24}$ und v_{34} (see picture on the right). There exist two left-orders that are induced by S with $m_1 \ S \ m_2 \ S \ m_3 \ S \ m_4$ and S^{-1}. We consider the left-order generated by S (the other one may by observed analog). For the relations $L_<$ and $R_<$ we find:

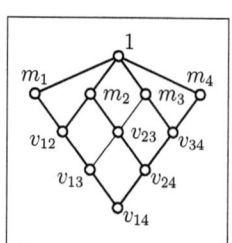

$L_<$	v_{14}	v_{13}	v_{12}	m_1	v_{24}	v_{23}	m_2	v_{34}	m_3	m_4	1
$R_<$	v_{14}	v_{24}	v_{34}	m_4	v_{13}	v_{23}	m_3	v_{12}	m_2	m_1	1

Hence, we yield for the diagram realizer l and r:

$$l(m_1) < l(v_{24}) = l(m_2) + l(m_3) + l(m_4)$$
$$r(m_4) < r(v_{13}) = r(m_1) + r(m_2) + r(m_3)$$

Recall that the equality sign holds in both expressions since we must preserve the attribute additivity convention.

Moreover, let the layers be lines of the form $c = \alpha \cdot x + \beta \cdot y$, which is reasonable due to our considerations in Section 5.3.2. Therefore, $\alpha, \beta > 0, c < 0$ and $\alpha \cdot l(m_i) + \beta \cdot r(m_i) = c$ for all $i \in \{1, 2, 3, 4\}$. We conclude as follows:

$$\alpha \cdot l(m_4) + \beta \cdot r(m_4) = c$$
$$\alpha \cdot l(m_4) + \beta \cdot (r(m_1) + r(m_2) + r(m_3)) > c$$
$$\alpha \cdot l(m_1) + \beta \cdot r(m_1) = c$$
$$\alpha \cdot (l(m_2) + l(m_3) + l(m_4)) + \beta \cdot r(m_1) > c$$
$$\alpha \cdot (l(m_2) + l(m_3) + 2l(m_4)) + \beta \cdot (2r(m_1) + r(m_2) + r(m_3)) > 2c$$
$$\alpha \cdot l(m_4) + \beta \cdot r(m_1) > 0$$

which is a contradiction since $\alpha \cdot l(m_4) < 0$ and $\beta \cdot r(m_1) < 0$.

We may subsume, that attribute additive diagrams may be drawn with the left-right-numbering-method. However, it is not possible to additionally put the $\bigwedge$-irreducibles on layers, as this is commonly done when drawing lattices (at least, if laid out manually).

Chapter 6

Left-relations on Contexts

In Chapter 4 we introduced left-relations on lattices. We noticed that we do not necessarily need the whole relation for deciding the planarity of the considered structure. In particular, the FPC allows such a characterization if we know L only on the set $M \times M$.

Now we want to observe whether it is similarly fruitful to restrict L to the set $J \times M$, where J and M are as usual the sets of $\bigvee$-irreducibles and $\bigwedge$-irreducibles respectively. This idea is obviously inspired by the FCA-viewpoint on lattice theory. The standard context of a lattice itself consists of the sets J and M and the incidence relation, also a subset of $J \times M$. Hence, we can just insert L and R into the empty cells of a respective cross table.

This may recall of the Ferrers-graphs introduced in Section 3.6. Intuitively, a (bare) Ferrers-graph is also defined on the set of empty cells (j, m) of a cross table meeting $j \parallel m$. The notation with two symbols L and R determines two vertex classes. Indeed, a left-order on a lattice already generates a *bipartite* Ferrers-graph on its standard context, as we have seen in Lemma 3.42.

We will in the following go the other way round. Based on a bipartite Ferrers-graph, whose vertex classes we are going to consider as a restricted left-relation, we want to construct the whole left-relation and show that it is a left-order indeed.

In fact, in the last chapters we did not yet give an efficient possibility to find a left-order on a lattice or to decide whether it exists at all. The most naive ways would be to test all conjugate relations of a lattice to be conjugate orders or all sorting relations to be left-orders. Although that may be satisfactory for small lattices, in general one may have up to $|M|!$ sorting relations, i.e. the process may become very time consuming.

However, since it is "easy" by means of time complexity to test whether a graph is bipartite, we can now specify a method (see Section 6.3) to check a given lattice for being planar and to provide a plane diagram, if possible, in polynomial time. We may even give the number of all plane diagrams up to similarity (see Section 6.3.2), although writing them all down can not be polynomial anymore.

6.1 Definition and First Properties

We decided to use the language of lattice theory here as well. Many results can be directly transferred to FCA-notions. In particular, we could have based our considerations on *contexts* in general instead of *standard contexts of lattices*, which are according to Theorem 3.36, the appropriate *reduced contexts*. These are the smallest representatives (in terms of the size of G and M) of their accompanying lattices. However, since we are interested not in contexts itself but rather in some properties (relying on planarity) of the appropriate lattices we decided to restrict to consider the reduced ones as canonical forms.

Definition 6.1 *[Zsc08] Let $\mathfrak{V}$ be a lattice and $\mathbb{K} = (J, M, \leq)$ its standard context.*

1. *The relation $\|_\mathbb{K} = (J \times M) \cap \|$ is called* incomparability relation *on $\mathbb{K}$.*

2. *A relation $\mathfrak{L} \subseteq \|_\mathbb{K}$ is called* left-relation *on $\mathbb{K}$. We denote $\mathfrak{R} := \|_\mathbb{K} \setminus \mathfrak{L}$.*

That definition may be surprising since it is not as intuitive as Definitions 4.2 and 5.2. There we developed the concept of "left of" out of a geometric view. Of course, one can understand it as a relation stating which $\bigvee$-irreducibles are left or right of which $\bigwedge$-irreducibles but this does not fit into a diagrammatic understanding of the lattice. However, in the following we will reinterpret vertex classes of bipartite Ferrers-graphs as left-relations of contexts. Instead of stating that "(j, m) is in vertex class F_1" we read "j is left of m". We will evidence that these particular relations can be transformed to left-orders, so the notation becomes reasonable.

In order to find such a left-order $\hat{L}$, we first construct its "core" $\tilde{L}$ out of $\mathfrak{L}$. After some modifications $\tilde{L}$ will play a role similar to a sorting relation (see Definition 4.1) for determining $\hat{L}$.

Definition 6.2 *[Zsc07a] Let $\mathfrak{V}$ be a lattice and $\mathbb{K} = (J, M, \leq)$ its standard context. Let $\mathfrak{L}$ and $\mathfrak{R}$ be disjoint subsets of $\overline{I}$. The relation*

$$\tilde{L} := \{(m, n) \in M \times M \mid m \| n \text{ and } (\exists j \in J : m \geq j \mathfrak{L} n \text{ or } n \geq j \mathfrak{R} m)\}.$$

is called induced relation *by $\mathfrak{L}$.* ◊

6.1 Definition and First Properties

See the picture on the right for an intuitive access to this notion. On the left you consider the case $j \leq m$ (j is below m), $j \, \tilde{\mathfrak{L}} \, n$ (j is left of n). Visually m is then left of n too. On the right, similarly $j \, \tilde{\mathfrak{R}} \, m$ (j is right of m or m is left of j) and $j \leq n$ suggest that m is again left of n. Of course, in this general setting both $m \, \tilde{L} \, n$ and $n \, \tilde{L} \, m$ may occur.

Now let $\mathbb{K} = (J, M, \leq)$ be the standard context of a lattice $\mathfrak{V}$ possessing the bipartite Ferrers-Graph $\tilde{\Gamma}$. Let the vertex classes of a bipartition be denoted by $\tilde{\mathfrak{L}}$ and $\tilde{\mathfrak{R}}$. Then $\tilde{\mathfrak{L}} \setminus \{(j,m) \mid j > m\}$ is a left-relation on $\mathbb{K}$, but not $\tilde{\mathfrak{L}}$ itself. Fortunately the set $\{(j,m) \mid j > m\}$ does not affect the bipartiteness of $\tilde{\Gamma}$ since it refers exactly to the isolated vertices of $\tilde{\Gamma}$:

Proposition 6.3 *[Zsc07a] Let $\mathbb{K} = (J, M, \leq)$ be the standard context of a lattice $\mathfrak{V}$ and $\tilde{\Gamma}$ its Ferrers-graph. A vertex (j,m) of $\tilde{\Gamma}$ is isolated if and only if $j > m$.*

Proof:
$\Leftarrow$: Let $j > m$. Furthermore let $h \in J$ and $n \in M$, s.t. $h \leq m$ and $g \leq n$ holds. We notice $h \leq m < j \leq n$, i.e. $h \leq n$. Therefore (h,n) is no vertex of $\tilde{\Gamma}$, i.e. there is no edge between (j,m) and (h,n).
$\Rightarrow$: Since $(j,m) \in V(\Gamma)$ we know $j \not\leq m$.

1. If there is no $\bigvee$-irreducible $h \in J$ satisfying $h \leq m$ then we conclude $m = 0_{\mathfrak{V}}$ and hence $j > m$.

2. If there is no $\bigwedge$-irreducible $n \in M$ satisfying $j \leq n$ then we conclude $j = 1_{\mathfrak{V}}$ and hence $j > m$.

3. Otherwise the sets j' and m' are non-empty. All objects $h \in m'$ and all attributes $n \in j'$ meet the condition $h \leq n$ since there is no edge between (j,m) and (h,n) in $\tilde{\Gamma}$. Hence, we conclude $j'' \supseteq m' \uplus \{j\}$, i.e. $m < j$.

$\square$

According to Proposition 6.3, the bipartition classes $\mathfrak{L}$ and $\mathfrak{R}$ of a bipartite *bare* Ferrers-graph Γ are exactly left-relations on $\mathbb{K}$. The left-relation $\mathfrak{L}$ and its partner $\mathfrak{R}$ have some additional properties. First, we easily observe that

Definition 6.2 becomes slightly simplified. For incomparable $\bigwedge$-irreducibles m and n the following equivalence holds:

$$\exists j \in J : m \geq j \; \mathfrak{L} \; n \iff \exists h \in J : n \geq h \; \mathfrak{R} \; m. \tag{6.1}$$

The next lemma states that $\tilde{L}$ contains "nearly" a sorting relation S. This sorting relation will be extended to a left-order in Lemma 6.5.

Lemma 6.4 *[Zsc07a] Let $\mathbb{K} = (J, M, \leq)$ be the standard context of a lattice $\mathfrak{V}$ and Γ its bipartite bare Ferrers-graph with vertex classes $\mathfrak{L}$ and $\mathfrak{R}$. Let $\tilde{L}$ be the relation induced by $\mathfrak{L}$. Then $\tilde{L}$ satisfies the following conditions:*

1. *$\tilde{L}$ is asymmetric,*

2. *$\tilde{L}$ is connex on pairs of incomparable $\bigwedge$-irreducible elements[1],*

3. *For any three $\bigwedge$-irreducibles fulfilling $m_1 \; \tilde{L} \; m_2 \; \tilde{L} \; m_3$ either $m_1 \; \tilde{L} \; m_3$ or $m_3 \; \tilde{L} \; m_1$ holds.*

Proof:

1. Let m and n be $\bigwedge$-irreducibles fulfilling $m \; \tilde{L} \; n$ and $n \; \tilde{L} \; m$. In a respective cross table one of the three cases depicted on the right occurs. The left and the right one contradict the fact that $\mathfrak{L}$ and $\mathfrak{R}$ respectively are vertex classes of Γ.

   Let us consider the case in the middle. There exists a $\bigvee$-irreducible $j \in J$ with $j \parallel m$ and $j \leq n$ (by definition of $\tilde{L}$ we know $m \parallel n$). Both $j \; \mathfrak{L} \; m$ and $j \; \mathfrak{R} \; m$ contradict the bipartition property. Hence, $\tilde{L}$ is asymmetric.

2. Connexity is obvious: For two incomparable $\bigwedge$-irreducibles $m \parallel n$ we find a $\bigvee$-irreducible j meeting $j \leq m$ and $j \parallel n$, i.e. $(j, n) \in \mathfrak{L} \cup \mathfrak{R}$.

3. Let $m_1 \leq m_3$. Then there exist $\bigvee$-irreducibles j_1 and j_2 satisfying $m_1 \geq j_1 \; \mathfrak{L} \; m_2$ and $m_2 \geq j_2 \; \mathfrak{L} \; m_3$. This contradicts Γ to be bipartite since $m_1 \leq m_3$ implies $j_1 \leq m_3$, i.e. $(j_1, m_2)E(j_2, m_3)$. The case $m_1 \geq m_3$ yields the same contradiction if we consider the relation $\mathfrak{R}$ instead.

   We conclude $m_1 \parallel m_3$. The claim follows with assertions 1. and 2. $\square$

---

[1] By this formulation, we mean $\tilde{L} \cup \tilde{L}^{-1} = \parallel$.

6.1 Definition and First Properties 75

Unfortunately, we could not prove $\tilde{L}$ to be transitive as well. If we just assume that condition then indeed $\tilde{L}$ can be extended to a left-order, as the following lemma suggests.

Lemma 6.5 *Let* $\mathbb{K} = (J, M, \leq)$ *be the standard context of a lattice* $\mathfrak{V}$ *and* Γ *be its bipartite bare Ferrers-graph possessing the vertex classes* $\mathfrak{L}$ *and* $\mathfrak{R}$. *Let* $\tilde{L}$ *be the relation induced by* $\mathfrak{L}$. *Then the following implication holds:*

$$\tilde{L} \text{ is transitive} \implies \mathfrak{V} \text{ is planar.}$$

Proof: By Lemma 6.4 we know that $\tilde{L}$ is asymmetric and connex on incomparable pairs of $\bigwedge$-irreducibles. Since $\tilde{L}$ is also transitive we conclude that $\tilde{L}$ is a strict order on sets of $\bigwedge$-irreducibles possessing the same upper cover. Hence, there exists a sorting relation $S \subseteq \tilde{L}$ (see Definition 4.1). Let L be the (unique) left-relation induced by S.

1. We show $\tilde{L} \subseteq L$. Let $m_1 \tilde{L} n_1$. According to Definition 4.2, we have to show that there exists a pair of $\bigwedge$-irreducibles $(m_2, n_2) \in M(m_1, n_1)$ with $m_2 \tilde{L} n_2$ since then $\tilde{L}$ can be constructed out of S.

   Since $m_1 \tilde{L} n_1$ there exists a $\bigvee$-irreducible j fulfilling $j \leq m_1$ and $j \mathfrak{L} n_1$. From $m_1 \leq m_2$ we conclude $j \leq m_2$. Consider a $\bigvee$-irreducible h satisfying $h \leq n_1$ and $h \parallel m_2$ (which exists because $n_1 \parallel m_2$). Then $h \leq n_2$ and since Γ is bipartite we conclude $h \mathfrak{R} m_2$, i.e. $m_2 \tilde{L} n_2$.

	m_1	n_1	m_2	n_2
j	×	$\mathfrak{L}$	×	
h		×	$\mathfrak{R}$	×

2. We show that L is a strict order. We know with Lemma 6.4 and 1. that $\tilde{L} = L \cap (M \times M)$. By applying the FPC we need to prove only

$$m_1 \tilde{L} m_2 \tilde{L} m_3 \implies m_2 > (m_1 \wedge m_3)$$

   Let $m_1 \tilde{L} m_2 \tilde{L} m_3$. According to Condition 6.1 there exist $\bigvee$-irreducibles j_1 and j_2 fulfilling $j_1 \leq m_2$, $j_2 \leq m_2$, $j_1 \mathfrak{L} m_1$ and $j_2 \mathfrak{R} m_3$.

   Let $j_3 \in J$ meet $j_3 \leq m_1 \wedge m_3$. (If j_3 does not exist, then $m_1 \wedge m_3$ is the bottom element of $\mathfrak{V}$ and m_2 trivially greater.) Now $j_3 > m_2$ contradicts our assertion $m_1 \parallel m_2$ and both $j_3 \mathfrak{L} m_2$ and $j_3 \mathfrak{R} m_2$ contradict $\mathfrak{L}$ and $\mathfrak{R}$ to be bipartition classes of Γ. We conclude $j_3 \leq m_2$, i.e. $m_2 > m_1 \wedge m_3$.

	m_1	m_2	m_3
j_1	$\mathfrak{L}$	×	
j_2		×	$\mathfrak{R}$
j_3	×	×	×

By the use of $\mathfrak{L}$ we constructed a left-order L on $\mathfrak{V}$. With Theorem 4.9 and Theorem 3.20 we conclude that $\mathfrak{V}$ is planar. □

With the previous Lemmas 6.4 and 6.5 we realize that we are "nearly finished": From the vertex classes $\mathfrak{L}$ and $\mathfrak{R}$ of the bipartite bare Ferrers-graph Γ we constructed via the induced relation $\tilde{L}$ a left-order L assuring that the underlying lattice $\mathfrak{V}$ is planar. See Figure 6.1 for an example. We are given a context $\mathbb{K}$ together with its Ferrers-graph Γ. It is easy to find the unique bipartition of that graph. From the respective vertex classes $\mathfrak{L}$ and $\mathfrak{R}$ one can calculate the induced relation $\tilde{L}$. In the first row one reads $n_3\ \tilde{L}\ n_2, n_4$ and from the second $n_1\ \tilde{L}\ n_2, n_4$. Indeed, by extending $\tilde{L}$ one gets a left-order that provides a plane diagram possessing the left-relation $\lambda \supseteq \tilde{L}$.

$\mathbb{K}$	n_1	n_2	n_3	n_4
j_1	×	×	$\mathfrak{R}$	×
j_2	×	$\mathfrak{L}$	×	$\mathfrak{L}$
j_3	$\mathfrak{R}$	×	$\mathfrak{R}$	×
j_4	×	$\mathfrak{L}$	×	×

$n_3\ \tilde{L}\ n_2$
$n_3\ \tilde{L}\ n_4$
$n_1\ \tilde{L}\ n_2$
$n_1\ \tilde{L}\ n_4$

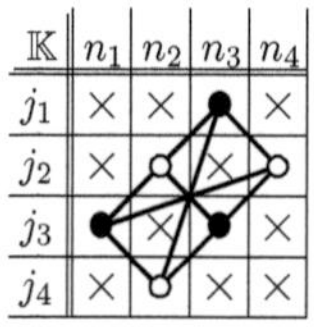

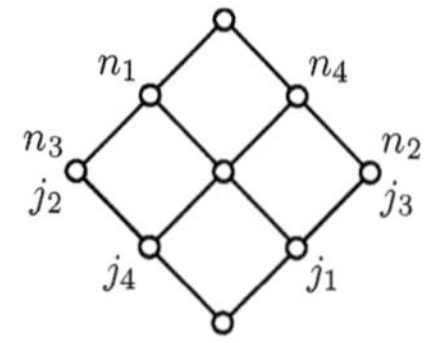

Figure 6.1: A reduced context $\mathbb{K}$ together with its bipartite Ferrers-graph (left), an appropriate bipartition (left middle), its induced relation (right middle) and a plane diagram drawn according to that induced relation (right).

The only assumption we had to take was the transitivity of $\tilde{L}$. Unfortunately it is not possible to prove this assertion as straightforward as the others. Let us consider the example in Figure 6.2. On the left you can see the standard context $\mathbb{K}$ of the lattice M_3. We notice that its Ferrers-graph (depicted in the middle) is bipartite and consists of three components. For the given choice of the bipartition classes, the induced relation $\tilde{L}$ is not transitive. However, we can "flip" for instance the component in the middle to make $\tilde{L}$ transitive. This will be our strategy in the following sections:

If $\tilde{L}$ is not transitive, Γ consists of at least three components which can be "turned around smartly". This will keep Γ bipartite and the induced relation becomes transitive.

$\mathbb{K}$	m_1	m_2	m_3
j_1	×	$\mathfrak{L}$	$\mathfrak{R}$
j_2	$\mathfrak{R}$	×	$\mathfrak{L}$
j_3	$\mathfrak{L}$	$\mathfrak{R}$	×

$m_2\ \tilde{L}\ m_3$
$m_3\ \tilde{L}\ m_1$
$m_1\ \tilde{L}\ m_2$

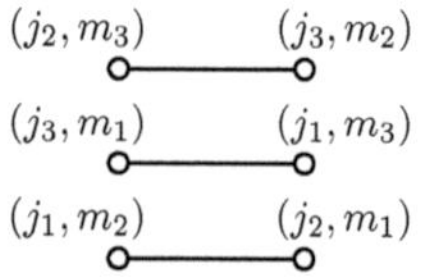

Figure 6.2: A construction of a the bipartition classes $\mathfrak{L}$ and $\mathfrak{R}$ of the Ferrers-graph Γ of the given context $\mathbb{K}$. The induced relation $\tilde{L}$ is not transitive.

6.2 The Components of Ferrers-graphs

We got a flavor of the general behavior of a bipartite Ferrers-graph Γ and its interrelation to a left-order on the underlying lattice $\mathfrak{V}$ in the last section. However, the example in Figure 6.2 suggests to further investigate the components of Γ in order to reach our aim, namely to generate left-orders of the underlying lattice out of Γ, if it is bipartite.

To gain a better visual understanding, we want to highlight how a component of the Ferrers-graph Γ looks like:
An edge sequence of Γ, as seen in a cross table of a context $\mathbb{K}$, corresponds to a pair of *fences*, i.e. alternate up and down leading edge sequences in a diagram of the respective lattice $\mathfrak{V}$. Vertices situated one upon the other are nodes of Γ and thus incomparable.

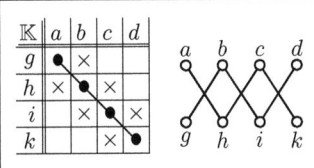

Normally we will refer to a component of Γ by an index, for instance $\Gamma_\circ$. Sometimes it will be helpful to describe a component by two $\bigwedge$-irreducibles "it lives on".

Definition 6.6 *[Zsc07a] Let $\mathbb{K} = (J, M, \leq)$ be a context and Γ its bare Ferrers-graph. Let $m, n \in M$ be incomparable. The (m,n)-component of Γ denoted by $\Gamma_{m,n}$ is that component of Γ containing all edges between m and n, i.e. all edges of the form $\{(j,m),(h,n)\}$ (for arbitrary elements $j, h \in J$).* ◊

Such components can easily be spotted in a cross table: $\Gamma_{m,n}$ is that component that contains all edges between the columns assigned to m and n respectively.

One might argue whether Definition 6.6 is well defined since edges of the form $\{(j,m),(h,n)\}$ may occur in several components of Γ. Fortunately this is not the case, as a simple observation shows: Let $(j_1,m)E(h_1,n)$ and $(j_2,m)E(h_2,n)$ be edges of Γ then we note by Definition 3.40 $h_1 \leq m$ and $j_2 \leq n$, i.e. both edges are connected in Γ via the edge $\{(h_1,n)(j_2,m)\}$ (see picture on the right). Hence, Γ possesses at most $\binom{|M|}{2}$ components.

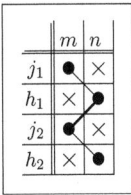

A component consists of pairs of $\bigvee$-irreducibles and $\bigwedge$-irreducibles. The $\bigwedge$-irreducible elements are in general not arbitrarily scattered in the underlying lattice. In fact it is quite fruitful to consider the interval where they are contained in.

78   **Chapter 6: Left-relations on Contexts**

Definition 6.7 *[Zsc08] Let $\mathfrak{V}$ be a lattice with Ferrers-graph Γ. Let $\Gamma_\circ$ be a component of Γ. By*

$$M(\Gamma_\circ) := \{m \in M \mid \exists j \in J : (j,m) \in V(\Gamma_\circ)\}$$

we denote the set of all $\bigwedge$-irreducibles which belong to a vertex of $\Gamma_\circ$. Dually we define the set $J(\Gamma_\circ)$. The interval $I(\Gamma_\circ) = [\underline{v}, \overline{v}]$ of a component $\Gamma_\circ$ is defined by

$$\underline{v} := \bigwedge M(\Gamma_\circ) \quad \text{and} \quad \overline{v} := \bigvee M(\Gamma_\circ).$$

$\Diamond$

First we show that indeed both $M(\Gamma_\circ)$ and $J(\Gamma_\circ)$ are contained within the interval of $\Gamma_\circ$, if $\Gamma_\circ$ consists of at least two elements.

Lemma 6.8 *Let $\mathfrak{V}$ be a lattice with bare Ferrers-graph Γ. Let $\Gamma_\circ$ be a component possessing the interval $[\underline{v}, \overline{v}] = I(\Gamma_\circ)$. Then*

$$M(\Gamma_\circ) \subseteq (\underline{v}, \overline{v}) \quad \text{and} \quad J(\Gamma_\circ) \subseteq (\underline{v}, \overline{v}).$$

Proof:

1. Let $m \in M(\Gamma_\circ)$. Since $\Gamma_\circ$ consists of at least two vertices, we find elements $n \in M(\Gamma_\circ)$ and $j, h \in J(\Gamma_\circ)$, s.t. $((j,m),(h,n))$ is an edge in $\Gamma_\circ$. By Definition 3.40 we note $m \parallel n$ and therefore $\underline{v} \leq m \wedge n < m < m \vee n \leq \overline{v}$.

2. Let $j \in J(\Gamma_\circ)$. Since $\Gamma_\circ$ consists of at least two vertices, we find elements $m, n \in M(\Gamma_\circ)$ and $h \in J(\Gamma_\circ)$, s.t. $((j,m),(h,n))$ is an edge in $\Gamma_\circ$. By Definition 3.40 we note $j \leq n$ and $j \parallel m$ and find with the first statement of this Lemma $j \leq n < \overline{v}$ and $\underline{v} < m$, i.e. $j \not\leq \underline{v}$.

   We finally have to prove $j \nparallel \underline{v}$. It is a basic result of lattice theory that $j = \bigwedge\{\tilde{n} \in M \mid m \geq j\}$. Hence we have to show that $\tilde{n} \geq j \implies \tilde{n} \geq \underline{v}$ for all $\tilde{n} \in M$.

   Let $\tilde{n}$ be an arbitrary $\bigwedge$-irreducible satisfying $j \leq \tilde{n}$. Since $\tilde{n} \not\leq m$ (that would imply $j \leq m$) we have either $\tilde{n} > m > \underline{v}$ or $\tilde{n} \parallel m$. The latter case implies the existence of a $\bigvee$-irreducible $\tilde{h}$ satisfying $\tilde{n} \parallel \tilde{h} \leq m$. Hence $\{(\tilde{h}, \tilde{n}), (j, m)\}$ is an edge of $\Gamma_\circ$. Therefore $\tilde{n} \in M(\Gamma_\circ)$ and hence $\tilde{n} > \underline{v}$.

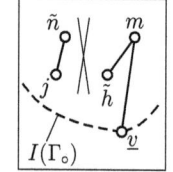

$\square$

In the remaining of this section we will tackle the following questions:

6.2 The Components of Ferrers-graphs

1. How does the induced relation look like on a component $\Gamma_\circ$? In particular, we will give evidence to the fact that the induced relation $\tilde{L}$, if restricted to one component (see Definition 6.13), is transitive, i.e. a strict order. That assures one assumption of our agenda, namely that by flipping over components of Γ one can indeed remove the "non-transitive" part of the induced relation.

2. Under which conditions are two components Γ_{m_1,n_1} and Γ_{m_2,n_2} equal? This is equivalent to the existence of an edge sequence

$$(j_1, m_1)E(h_1, n_1)E \ldots E(h_2, n_2)E(j_2, m_2) \text{ or}$$
$$(j_1, m_1)E(h_1, n_1)E \ldots E(j_2, m_2)E(h_2, n_2)$$

in the bare Ferrers-graph Γ of the respective context $\mathbb{K} = (J, M, \leq)$ (where $m_1, n_1, m_2, n_2 \in M$ and $j_1, h_1, j_2, h_2 \in J$).

3. Under which conditions is the $\bigwedge$-irreducible m an element of $M(\Gamma_{m_1,m_2})$? This is equivalent to the existence of an edge sequence

$$(j, m)E \ldots E(j_1, m_1)E(j_2, m_2)$$

in the bare Ferrers-graph Γ of the respective context $\mathbb{K} = (J, M, \leq)$ (where $m, m_1, m_2 \in M$ and $j, j_1, j_2 \in J$).

6.2.1 Connections in Intervals

First let us have a look back to the lattice perspective. We can also introduce the notion of a *connection* there. This concept, as well as the similar and more popular one of a *fence* (see [KR75, DP02]) denotes a path of a diagram of a lattice consisting of alternately upward and downward edge sequences. Contrary to other approaches, we allow as "endpoints" only $\bigvee$-irreducibles and $\bigwedge$-irreducibles since they play a prominent role in the appropriate standard context $\mathbb{K} = (J, M, \leq)$. Additionally, we consider these structures in certain intervals $[\underline{v}, \overline{v}]$. We will see later on, why this is reasonable.

Definition 6.9 *[Zsc07a] Let $\mathbb{K} = (J, M, \leq)$ be the the standard context of a lattice $\mathfrak{V}$ and $[\underline{v}, \overline{v}]$ be an interval in $\mathfrak{V}$. A sequence*

$$p = m_0 \geq j_1 \leq m_1 \geq j_2 \ldots \geq j_r \leq m_r$$

of $\bigvee$-irreducibles $j_i, i \in \{1, \ldots, r\}$ and $\bigwedge$-irreducibles $m_i, i \in \{0, \ldots, r\}$ is called connection *of m_0 and m_r in $[\underline{v}, \overline{v}]$ if*

$$\forall i \in \{0, \ldots, r\} : m_i \not\geq \overline{v} \quad \text{and} \quad \forall i \in \{1, \ldots, r\} : j_i \not\leq \underline{v}$$

If the condition $j_i \leq m_k \implies k \in \{i, i-1\}$ holds for all $i \in \{1, \ldots, r\}$ as well then p is called shortest connection.

Chapter 6: Left-relations on Contexts

See Figure 6.3 for a visualization of how a connection in $[\underline{v}, \overline{v}]$ looks like. A connection of m and n in $[\underline{v}, \overline{v}]$ is a set of directed edge sequences alternately running "down" from a $\bigwedge$-irreducible to a $\bigvee$-irreducible and "up" vice versa (see left diagram). These edges may leave the interval $[\underline{v}, \overline{v}]$ itself but may not enter the principal filter $[\overline{v})$ nor the principal ideal $(\underline{v}]$. There is another access to this definition: The context $\mathbb{K} = (J, M, \leq)$ itself is a directed acyclic bipartite graph consisting of the vertex classes J and M and the incidence relation $\leq$. In this setting (see right diagram) a connection is a sequence of edges running alternately down and up. Thereby the subsets $[\overline{v}) \cap M$ and $(\underline{v}] \cap J$ may not be included.

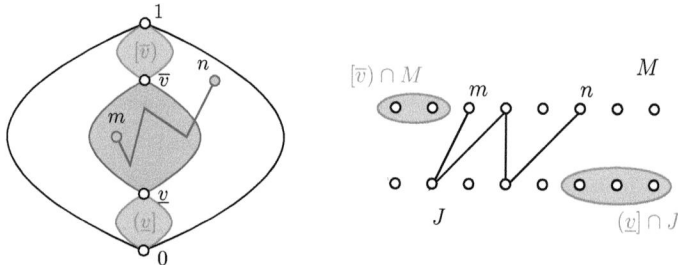

Figure 6.3: Two graphical descriptions of a *connection* between two $\bigwedge$-irreducibles m and n in an interval $[\underline{v}, \overline{v}]$.

One can easily verify that every connection p between n_0 and n_r contains a shortest connection q since we can construct q in the following way: Start with $q := n_0$. If n_i is the last element of the sequence of q, search (in p) for the highest index t with $h_t \leq n_i$ and add h_t to q. If h_i is the last element of q, search (in p) for the highest index t with $h_i \leq n_t$ and add n_t to q. Stop after adding n_r.

A shortest connection between n_0 and n_r is represented by a subcontext as depicted by a cross table on the right. The main diagonal and the secondary diagonal above are filled with crosses, while all the other cells are empty. Note that the sets of of empty cells marked with • and ○ respectively, are Ferrers-relations.

	n_0	n_1	...	n_r
h_1	×	×	• ... •	•
h_2	○	×	×	•
⋮			⋱	⋮
h_{r-1}	○	○		×	•
h_r	○	○		×	×

6.2 The Components of Ferrers-graphs

It is easy to prove that every two of these empty cells are in the same component of Γ since a vertex $(h_i, n_j) \in V(\Gamma(I))$ with $i < j$ has the neighbors

$$N(h_i, n_j) = \{(h_j, n_{i-1}), (h_j, n_i), (h_{j+1}, n_{i-1}), (h_{j+1}, n_i)\} \cap V(\Gamma)$$

In particular, if $n_0 \, \tilde{L} \, n_r$ then we conclude
$\begin{cases} h_i \, \mathfrak{R} \, n_j, & j < i, \\ h_i I n_j, & j \in \{i, i+1\}, \\ h_i \, \mathfrak{L} \, n_j, & j > i+1. \end{cases}$

In the following definition, we will specify the different kinds of connections more precisely. There are connections leading through the interval, connections inside the interval and finally triples of $\bigwedge$-irreducibles non-connected neither to each other nor the outside.

Definition 6.10 *[Zsc07a] Let $[\underline{v}, \overline{v}]$ be an interval in the lattice $\mathfrak{V}$.*

1. *A $\bigwedge$-irreducible $m \in [\underline{v}, \overline{v}]$ is called* bound *if there exists a $\bigwedge$-irreducible $n \notin [\underline{v}, \overline{v}]$ and a connection $p = m \ldots n$ in $[\underline{v}, \overline{v}]$.*

2. *Two $\bigwedge$-irreducibles $m, n \in [\underline{v}, \overline{v}]$ are called* connected *if there exists a connection $p = m \ldots n$ in $[\underline{v}, \overline{v}]$.*

3. *Three pairwise incomparable $\bigwedge$-irreducibles m_1, m_2 and m_3 are called* free triple *if none is bound and no two are connected in the interval $[m_1 \wedge m_2 \wedge m_3, m_1 \vee m_2 \vee m_3]$.*

In Figure 6.4 one can see how to imagine bounded and connected elements w.r.t. an interval $[\underline{v}, \overline{v}]$ and free triples in a diagram of a lattice: In the left picture m is bound since there is a connection to a $\bigwedge$-irreducible n which is not contained in $[\underline{v}, \overline{v}]$. In the middle m and n are connected. On the right the three $\bigwedge$-irreducibles m_1, m_2 and m_3 are a free triple since there are no edges between the three branches containing them.

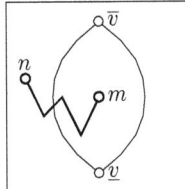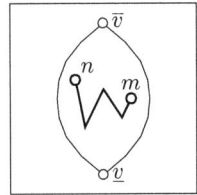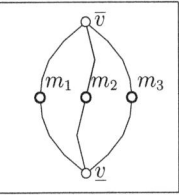

Figure 6.4: Examples for a bound element m (left), two connected elements m and n (middle) and a free triple m_1, m_2, m_3 (right).

Chapter 6: Left-relations on Contexts

In the following remark we collect some basic properties of the relations bound and connected.

Remark 6.11 Let $\mathfrak{V}$ be a lattice, $m, n \in M(\mathfrak{V})$ and I be an interval in $\mathfrak{V}$. The following properties hold:

1. If m is bound in I and m, n are connected in I then n is bound in I.
2. If $I \supsetneq [m \wedge n, m \vee n]$ then m, n are connected in I.
3. Let $I' \supseteq I$ be another interval in $\mathfrak{V}$. If m, n are connected in I then m, n are connected in I'.
4. "connected in I" is a symmetric and transitive binary relation on $M(\mathfrak{V})$.

The proofs are easy and straightforward and will be omitted therefore.

Finally we will introduce a notion for the set of connected $\bigwedge$-irreducibles w.r.t. a given interval. It will be helpful later on, in particular when we distinguish between components of the Ferrers-graph of different *types* (see Subsection 6.2.4).

Definition 6.12 *[Zsc07a]* Let $[\underline{v}, \overline{v}]$ be an interval in the lattice $\mathfrak{V}$. Let $U[\underline{v}, \overline{v}]$ denote the set of non-bound $\bigwedge$-irreducibles in $[\underline{v}, \overline{v}]$. Let $m \in U[\underline{v}, \overline{v}]$. The set

$$U_m([\underline{v}, \overline{v}]) := \{n \in M \mid m \text{ and } n \text{ are connected}\}$$

is called the m-component of $[\underline{v}, \overline{v}]$. ◇

Obviously, "connected in $[\underline{v}, \overline{v}]$" is an equivalence relation on $U([\underline{v}, \overline{v}])$. Therefore, the set of m-components in $[\underline{v}, \overline{v}]$ is a partition of $U([\underline{v}, \overline{v}])$. We may also give a graph theoretic access of this notion: $U_m([\underline{v}, \overline{v}])$ consists of the $\bigwedge$-irreducibles of the components inbetween $[\underline{v}, \overline{v}]$ of the undirected graph of a lattice $(\mathfrak{V}, \leq)$ minus the vertices $\underline{v}$ and $\overline{v}$, i.e. of $(\mathfrak{V} \setminus \{\underline{v}, \overline{v}\}, \prec \cup \succ)$.

A pair of connections

$$p = (n_1 \geq) h_0 \leq n_1 \geq h_2 \leq \ldots \leq n_r$$
$$q = n_0 \geq h_1 \leq n_2 \geq \ldots \geq h_r (\leq n_r)$$

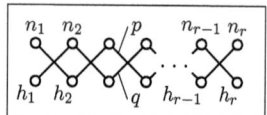

in an interval determines an edge sequence

$$S = (h_0, n_0) E(h_1, n_1) E(h_2, n_2) E \ldots E(h_r, n_r)$$

if $h_k \parallel n_k$ holds for all $k \in \{0, \ldots, r\}$. This principle is very helpful for constructing edge sequences and will be used widely in the proofs of the following subsection. In Subsection 6.2.3 we will further investigate the interrelations between the two concepts.

6.2.2 Induced Relations on Components of Γ

Our efforts to construct a left-order on a lattice out of a bipartition $\mathfrak{L} \stackrel{.}{\cup} \mathfrak{R}$ of its Ferrers-graph Γ failed due to the fact that the induced relation $\tilde{L}$ is not inevitably transitive. We already gave the hint that $\tilde{L}$ does not meet our desires only if *several*, more precisely at least three, components of Γ are involved. We will prove that in this subsection and conclude that the fraction of $\tilde{L}$ determined by a component of Γ is indeed a strict order.

First, we have to further specify the language we use, in particular we want to be able to assign the vertex classes $\mathfrak{L}$ and $\mathfrak{R}$ of the bipartite Ferrers-graph to its components.

Definition 6.13 *[Zsc07a] Let $\mathbb{K} = (J, M, \leq)$ be the standard context of a lattice $\mathfrak{V}$ and its bare Ferrers-graph Γ be bipartite with vertex classes $\mathfrak{L}$ and $\mathfrak{R}$. Let $\{\Gamma_k \mid k = \{1, \ldots, n\}\}$ be the set of components of Γ. We partition the vertex classes into $\mathfrak{L}_k := \mathfrak{L} \cap V(\Gamma_k)$ and $\mathfrak{R}_k := \mathfrak{R} \cap V(\Gamma_k)$.*
Furthermore let $\{\tilde{L}_i \mid j \in \{1, \ldots, n\}\}$ be binary relations on M defined by

$$m \, \tilde{L}_k \, n : \iff m \parallel n \text{ and } (\exists j \in J : m \geq j \, \mathfrak{L}_k \, n \text{ or } n \geq j \, \mathfrak{R}_k \, m).$$

Then we call $\{\tilde{L}_k \mid k \in \{1, \ldots, n\}\}$ the set of induced relations by $\mathfrak{L}$ and Γ. ◊

Obviously, $\bigcup \{\tilde{L}_k \mid k \in \{1, \ldots, n\}\} = \tilde{L}$ if $\tilde{L}$ is the induced relation by $\mathfrak{L}$. Since Γ is bipartite, also the following property is an immediate consequence:

$$\exists j \in J : m \geq j \, \mathfrak{L}_k \, n \iff \exists h \in J : n \geq h \, \mathfrak{R}_k \, m.$$

The introduction of the induced relations $\tilde{L}_k$ allows us to declare more specifically which component of Γ causes a $\bigwedge$-irreducible to be left or right of another one. Recall that this is uniquely determined since all edges between two columns of the respective cross table are in the same component. See Figure 6.5 for an example. The lattice similar to M_3 on the left has four $\bigwedge$-irreducibles n_1, n_2, n_3 and n_4 and four $\bigvee$-irreducibles j_1, j_2, j_3 and j_4. One can easily encounter a bipartition of its Ferrers-graph into the two classes $\mathfrak{L}$ and $\mathfrak{R}$. The empty cell (j_4, n_3) will not be of interest since $j_4 > n_3$, i.e. this node is isolated due to Proposition 6.3. Another partition of the vertex set of Γ is given by its decomposition into its components Γ_1 (upper left), Γ_2 (middle) and Γ_3 (lower right). Combining both partitions one gets a new partition with six classes $\mathfrak{L}_1, \ldots, \mathfrak{R}_3$ providing the induced relations given on the right.

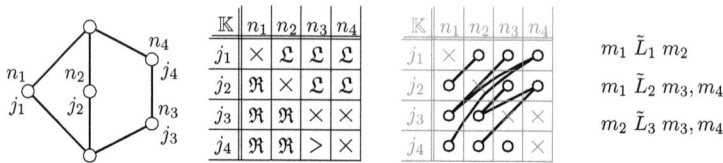

Figure 6.5: A lattice (left, given by a diagram) together with a bipartition of the Ferrers-graph of its standard context and a partition into its components (middle) and the resulting induced relations (right).

The following two lemmas are very technical. Their essence is the following (see picture and compare Lemma 6.14): connections between $\bigwedge$-irreducibles reduce the number of the respective components of Γ and allow further assertions about the induced relation. Sloppily spoken, connected elements in the interval of a Ferrers-graph do not harm the transitivity of the induced relation of that component.

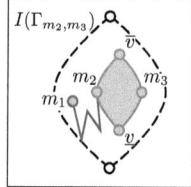

Lemma 6.14 *[Zsc07a, Zsc08]* Let $\mathbb{K} = (J, M, \leq)$ be the standard context of a lattice $\mathfrak{V}$ and Γ its Ferrers-graph. Let $m_1, m_2, m_3 \in M$ with $m_2 \parallel m_3$, s.t. there is a connection between m_1 and m_2 in $[\underline{v}, \overline{v}] := [m_2 \wedge m_3, m_2 \vee m_3]$.

If m_1, m_3 and m_2, m_3 are not connected in $[\underline{v}, \overline{v}]$ then $\Gamma_{m_2,m_3} = \Gamma_{m_1,m_3}$. Otherwise (at least) one of the following equalities holds: $\Gamma_{m_2,m_3} = \Gamma_{m_1,m_2}$ or $\Gamma_{m_2,m_3} = \Gamma_{m_1,m_3}$.

Let furthermore Γ be bipartite, s.t. $\tilde{L}_{ij}$ are the induced relations of Γ_{m_i,m_j} ($i \neq j \in \{1,2,3\}$). If both m_1, m_3 and m_2, m_3 are not connected in $[\underline{v}, \overline{v}]$ then $m_2 \tilde{L}_{23} m_3 \iff m_1 \tilde{L}_{23} m_3$. Otherwise $m_2 \tilde{L}_{23} m_3$ holds if and only if $m_1 \tilde{L}_{23} m_3$ or $m_2 \tilde{L}_{23} m_1$.

Proof: Let $q = n_0 \geq h_1 \ldots \geq h_r \leq n_r$ be a shortest connection in $[\underline{v}, \overline{v}]$ for adequate chosen $\bigvee$-irreducibles h_i and $\bigwedge$-irreducibles n_i between $n_0 := m_1$ and $n_r := m_2$. Let $\{(j_1, m_2), (j_2, m_1)\}$ be an edge of Γ_{m_1,m_2}. We distinguish the following cases:

1. Let $j_3 \not\leq n_i$ for all $i \in \{0, \ldots, r\}$ and $h_i \not\leq m_3$ for all $i \in \{1, \ldots, r\}$. This is, for instance, the case if m_1, m_3 and m_2, m_3 are not connected in $[\underline{v}, \overline{v}]$. Then

$$S_1 := (j_3, n_0) E(h_1, m_3) E(j_3, n_1) E(h_2, m_3) E \ldots E(h_r, m_3) E(j_3, n_r),$$

6.2 The Components of Ferrers-graphs

is an edge sequence and hence $\Gamma_{m_2,m_3} = \Gamma_{m_1,m_3}$. If additionally Γ is bipartite we conclude

$$m_2 \,\tilde{L}_{23}\, m_3 \implies j_2 \,\mathfrak{L}_{23}\, m_3 \implies j_3 \,\mathfrak{R}_{23}\, n_r \implies$$
$$\ldots \implies j_3 \,\mathfrak{R}_{23}\, n_0 \implies m_1 \,\tilde{L}_{23}\, m_3,$$

and, in particular, $\tilde{L}_{13} = \tilde{L}_{23}$. See Figure 6.6 for a visualization of this argument both on the lattice and the context side.

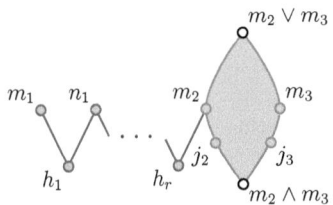 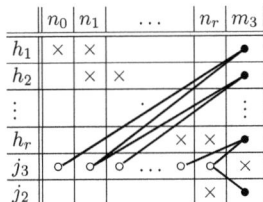

Figure 6.6: How to find the edge sequence S_1 in a lattice (diagram) and in the appropriate standard context. The black nodes • symbolize the respective cell to be in vertex class $\mathfrak{L}$ and the white ones ○ in class $\mathfrak{R}$.

2. Let $j_3 \leq n_k$ for one $k \in \{0, \ldots, r\}$ and $h_r \not\leq n_k$. Then $h_r < n_r$ implies $k < r$. Then

$$(j_2, m_3) E(j_3, n_r) E(j_2, n_k) E(h_k, n_r) E(h_r, n_{k-1}) E(h_{k-1}, n_r) E$$
$$\ldots E(h_2, n_r) E(h_r, n_0)$$

is an edge sequence and hence $\Gamma_{m_2,m_3} = \Gamma_{m_1,m_2}$. Additionally, if Γ is bipartite then $m_2 \,\tilde{L}_{23}\, m_3 \iff m_2 \,\tilde{L}_{23}\, m_1$ (see Figure 6.7).

3. Let $h_k \leq m_3$ for one $k \in \{1, \ldots, r\}$. From $h_k \not\leq m_2 \wedge m_3$ we conclude $k < r$. Then

$$(j_3, m_2) E(j_2, m_3) E(h_k, n_r) E(h_r, n_{k-1}) E(h_{k-1}, n_r) E \ldots E(h_1, n_r) E(h_r, n_0)$$

is an edge sequence and hence $\Gamma_{m_2,m_3} = \Gamma_{m_1,m_2}$. Additionally, if Γ is bipartite then $m_2 \,\tilde{L}_{23}\, m_3 \iff m_2 \,\tilde{L}_{23}\, m_1$ (see Figure 6.7).

4. Finally, let k be an index, s.t. $j_3, h_r \leq n_k$. Note $k < r$, since $(j_3, m_2) \in V(\Gamma)$ and hence $j_3 \parallel m_2$. Since q is a shortest connection, we observe $k = r - 1$. Furthermore $n_{r-1} \not\geq m_2 \vee m_3$ since q is a connection in $[\underline{v}, \overline{v}]$,

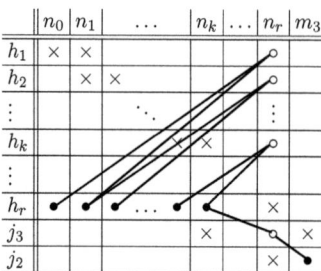 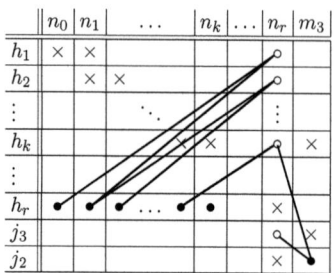

Figure 6.7: Visualization of the cases 2 (left) and 3 (right) of the proof.

hence $n_{r-1} \not\geq m_2$ (or $n_{r-1} \not\geq m_3$). Therefore we find a $\bigvee$-irreducible j satisfying $j \leq m_2$ (or $j \leq m_3$ respectively) and $j \not\leq n_{r-1}$. Then

$(j_2, m_3)E(j_3, m_2)E(j, n_{r-1})E(h_{r-1}, n_r)E(h_r, n_{r-1})E \ldots E(h_r, n_0)E(h_1, n_r)$

or

$(j_2, m_3)E(j_3, m_2)E(h_r, m_2)E(j, n_{r-1})E(h_{r-1}, m_3)E \ldots E(h_1, m_3)E(j_3, n_0)$

is an edge sequence (Recall that $j_3 \not\leq n_i$ holds for all $i \in \{0, \ldots, r\}$ and $h_i \not\leq m_3$ for all $i \in \{1, \ldots, r\}$, otherwise case 2 or 3 respectively can be applied.) We conclude $\Gamma_{m_2, m_3} = \Gamma_{m_1, m_2}$ (or $\Gamma_{m_2, m_3} = \Gamma_{m_1, m_3}$ respectively). If Γ is bipartite, then we additionally find $m_2 \ \tilde{L}_{23} \ m_3 \iff m_2 \ \tilde{L}_{23} \ m_1$ (or $m_2 \ \tilde{L}_{23} \ m_3 \iff m_1 \ \tilde{L}_{23} \ m_3$ respectively). □

Lemma 6.15 *[Zsc07a] Let $\mathbb{K} = (J, M, \leq)$ be the standard context of a lattice $\mathfrak{V}$ and Γ its bipartite Ferrers-graph. Let $m_1, m_2, m_3 \in M$ be pairwise incomparable, s.t. m_1, m_2 or m_2, m_3 are connected in the interval*

$$[\underline{v}, \overline{v}] := [m_1 \wedge m_2 \wedge m_3, m_1 \vee m_2 \vee m_3].$$

Then $m_1 \ \tilde{L}_{12} \ m_2 \ \tilde{L}_{23} \ m_3 \implies m_1 \ \tilde{L}_* \ m_3$ with $\tilde{L}_* \in \{\tilde{L}_{12}, \tilde{L}_{23}\}$.

Proof: We denote $v_{12} := m_1 \wedge m_2$, $v_{23} := m_2 \wedge m_3$, $v^{12} := m_1 \vee m_2$ and $v^{23} := m_2 \vee m_3$.

1. Let $v_{12} \not\leq v_{23}$. Then we find a $\bigvee$-irreducible j satisfying $v_{12} \geq j \not\leq v_{23}$. Hence m_1 and m_2 are connected via $p = m_1 \geq j \leq m_2$ in $[v_{23}, v^{23}]$. We may apply Lemma 6.14 and conclude $m_1 \ \tilde{L}_{23} \ m_3$ (recall that $\tilde{L}$ is antisymmetric due to Lemma 6.4).
   In analogy we find $v_{12} \not\geq v_{23} \implies m_1 \ \tilde{L}_{12} \ m_3$.

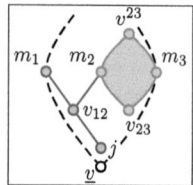

6.2 The Components of Ferrers-graphs

2. Let $v^{12} \not\leq v^{23}$. Then we find a $\bigwedge$-irreducible m satisfying $v^{23} \leq m \not\geq v^{12}$. Hence m_2 and m_3 are connected via $p = m_2 \geq j_2 \leq m \geq j_3 \leq m_3$ (with appropriate $j_2, j_3 \in J$) in $[v_{12}, v^{12}]$. We may again apply Lemma 6.14 and conclude $m_1 \tilde{L}_{12} m_3$.

   In analogy we find that $v_{12} \not\geq v_{23}$ implies $m_1 \tilde{L}_{23} m_3$.

3. Otherwise $[\underline{v}, \overline{v}] = [v_{23}, v^{23}] = [v_{12}, v^{12}]$. Lemma 6.14 yields in both cases $m_1 \tilde{L}_{23} m_3$. □

We figured out with Lemma 6.15 that three $\bigwedge$-irreducibles m_1, m_2, m_3 do not infringe the transitivity of the induced relation $\tilde{L}$ if two of them are connected. Now we investigate the case that one of them is bound. Again, this suffices $\tilde{L}$ to be transitive if restricted onto $\{m_1, m_2, m_3\}$.

Lemma 6.16 *[Zsc07a] Let* $\mathbb{K} = (J, M, \leq)$ *be the standard context of a lattice* $\mathfrak{V}$ *and* Γ *its bipartite Ferrers-graph. Let* $m_1, m_2, m_3 \in M$ *be pairwise incomparable, s.t.* m_1 *is bound and* m_1, m_2 *and* m_1, m_3 *are not connected in the interval* $[\underline{v}, \overline{v}] := [m_1 \wedge m_2 \wedge m_3, m_1 \vee m_2 \vee m_3]$. *Then* $\Gamma_{m_1, m_2} = \Gamma_{m_1, m_3}$ *follows and in particular* $m_1 \tilde{L}_{12} m_2 \iff m_1 \tilde{L}_{12} m_3$.

Proof: Let $p = n_0 \geq h_1 \leq n_1 \geq \ldots \leq n_{r-1} \geq h_r \leq n_r$ be a shortest connection between $n_0 := m_1$ and n_r with $n_r \notin [\underline{v}, \overline{v}]$. Let t be the biggest index with $n_t \in [\underline{v}, \overline{v}]$. We show first that $n_t \tilde{L} m_2 \iff n_t \tilde{L} m_3$.
According to Definition 6.9 we know $h_{t+1} \leq n_t$ and $h_{t+1} \leq n_{t+1}$. Since m_1, m_2 and m_1, m_3 are not connected we conclude $h_{t+1} \not\leq m_2$ and $h_{t+1} \not\leq m_3$.
Since m_1 is incomparable to m_2 and m_3, we find $\bigvee$-irreducibles j_2 and j_3 meeting the conditions $j_2 \leq m_2, j_3 \leq m_3$ and $j_2, j_3 \not\leq m_1$. Since m_1, m_2 and m_1, m_3 are not connected we notice $j_2, j_3 \not\leq n_t, n_{t+1}$. Now we distinguish following cases:

1. $n_{t+1} \not\geq \underline{v}$: We find a $\bigvee$-irreducible $h \leq \underline{v}$ not incident with n_{t+1} since $n_{t+1} \notin Int(\underline{v})$. Therefore the following edge sequence is in Γ (see Figure 6.8):
$$(j_1, n_t)E(h_{t+1}, m_1)E(h, n_{t+1})E(h_{t+1}, m_3)E(j_3, n_t).$$
Since Γ is bipartite we conclude $n_t \tilde{L} m_2 \iff n_t \tilde{L} m_3$.

2. $n_{t+1} \geq \underline{v}$: We note $n_{t+1} \parallel \overline{v}$ since we assumed $n_{t+1} \notin [\underline{v}, \overline{v}]$. Hence we find a $\bigwedge$-irreducible $\tilde{m}$ satisfying $\tilde{m} \parallel n_{t+1}$. Finally there exists a $\bigvee$-irreducible h with $h \leq n_{t+1}$ and $h \not\leq \tilde{m}$. Therefore we find the following edge sequence in $\Gamma(I)$ (see Figure 6.9):
$$(j_2, n_t)E(h_{t+1}, m_2)E(j_2, n_{t+1})E(h, \tilde{m})E(j_3, n_{t+1})E(h_{t+1}, m_3)E(j_3, n_t).$$
Since Γ is bipartite we conclude $n_t \tilde{L} m_2 \iff n_t \tilde{L} m_3$.

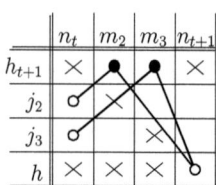 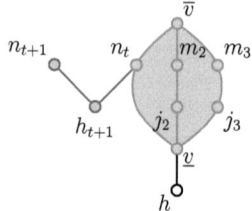

Figure 6.8: An illustration for the proof (case 1.). The filled dots • symbolize $\mathfrak{L}$ and the circles ○ symbolize $\mathfrak{R}$.

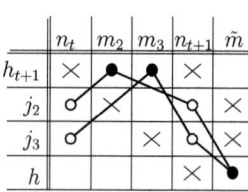 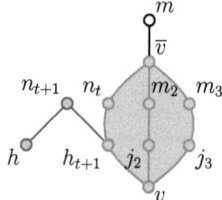

Figure 6.9: An illustration for the proof (case 2.). The filled dots • symbolize $\mathfrak{L}$ and the circles ○ symbolize $\mathfrak{R}$.

Since n_t and m_1 are connected, we finally apply Lemma 6.14 and conclude

$$m_1 \, \tilde{L}_{12} \, m_2 \iff n_t \, \tilde{L}_{12} \, m_2 \iff n_t \, \tilde{L}_{12} \, m_3 \iff m_1 \, \tilde{L}_{12} \, m_3.$$

Recall that $n_t \in [\underline{v}, \overline{v}]$, i.e. $[n_t \wedge m_2, n_t \vee m_2] \subseteq [\underline{v}, \overline{v}]$. Hence m_1, m_2 and m_1, m_3 are not connected in that interval as well. The same holds for the interval $[n_t \wedge m_3, n_t \vee m_3]$.) □

We mentioned already that the free triples will turn out to be the only problematic cases that may violate the transitivity of $\tilde{L}$. Now we are able to prove that:

Corollary 6.17 *[Zsc07a] Let $\mathbb{K} = (J, M, \leq)$ be the standard context of a lattice $\mathfrak{V}$ and Γ its bipartite Ferrers-graph. Let $m_1, m_2, m_3 \in M$. If $m_1 \, \tilde{L} \, m_2 \, \tilde{L} \, m_3 \, \tilde{L} \, m_1$ then (m_1, m_2, m_3) is a free triple.*

Proof: : Let $\{i, j, k\} = \{1, 2, 3\}$.

1. No two $\bigwedge$-irreducibles m_i, m_j are connected since $m_i \, \tilde{L} \, m_j \, \tilde{L} \, m_k \implies m_i \, \tilde{L} \, m_k$ (see Lemma 6.15).

6.2 The Components of Ferrers-graphs

2. No $\bigwedge$-irreducible m_i is bound since then $m_i \; \tilde{L} \; m_j \; \tilde{L} \; m_k \implies m_i \; \tilde{L} \; m_k$ (see Lemma 6.16).

By Definition 6.10 (m_1, m_2, m_3) is then a free triple. □

We want to show that components $\Gamma_\circ$ of the bipartite Ferrers-graph are "nice", i.e. their induced relation a strict order. We observed now that the only obstacle on our way is that of free triples. Now we want to exclude free triples to be in one of such component (i.e. in the set $M(\Gamma_\circ)$). The three components defined by a free triple are even all disjoint, as we see now:

Lemma 6.18 *[Zsc07a] Let $\mathbb{K} = (J, M, \leq)$ be the standard context of a lattice $\mathfrak{V}$ and Γ its bipartite Ferrers-graph. Let $m_1, m_2, m_3 \in M$. Then (m_1, m_2, m_3) is a free triple if and only if the components Γ_{m_i,m_j} ($i \neq j \in \{1, 2, 3\}$) are pairwise disjoint.*

Proof: $\Rightarrow$: We suppose $\Gamma_{m_1,m_2} = \Gamma_{m_1,m_3}$. W.l.o.g. we find an edge sequence

$$(j_1, m_1)E(j_2, m_2)E(h_0, n_0)E(h_1, n_1)E \ldots E(h_r, n_r)E(j_3, m_3)E(j_4, m_1).$$

in Γ (with appropriate $\bigvee$-irreducibles $j_1, \ldots, j_4, h_0, \ldots, h_r$ and $\bigwedge$-irreducibles $n_0, \ldots, n_r$. Since none of m_1, m_2 and m_3 is bound we know that

$$\{j_1, \ldots, j_4, h_0, \ldots, h_r, n_0, \ldots, n_r\} \subseteq [\underline{v}, \overline{v}]$$

(see Definition 6.10). Hence $p - m_2, h_0, \ldots, m$ is a connection in the interval $[m_1 \wedge m_2 \wedge m_3, m_1 \vee m_2 \vee m_3]$ (see Definition 6.9) where $m \in \{m_1, m_3\}$. This contradicts our assertion that no two attributes are connected.

$\Leftarrow$: If (m_1, m_2, m_3) is not a free triple then two $\bigwedge$-irreducibles are connected or one is bound. In both cases we find with Lemma 6.15 and Lemma 6.16 respectively that not all components Γ_{m_i,m_j} are disjoint. □

We can subsume the last two results to the following:

Proposition 6.19 *Let $\mathbb{K} = (J, M, \leq)$ be the standard context of a lattice $\mathfrak{V}$ and $\mathfrak{L} \cup \mathfrak{R}$ be a bipartition of its Ferrers-graph Γ. If the relation $\tilde{L}$ induced by $\mathfrak{L}$ is not transitive then Γ consists of at least three components.*

Proof: Let $m_1, m_2, m_3 \in M$ meeting $m_1 \; \tilde{L} \; m_2 \; \tilde{L} \; m_3$ and $(m_1, m_3) \notin \tilde{L}$. By Lemma 6.4 we know $m_3 \; \tilde{L} \; m_1$. By Corollary 6.17 we conclude (m_1, m_2, m_3) to be a free triple. With Lemma 6.18 we know that there exist at least three components of Γ. □

Now we have all ingredients for the final result of this subsection. Inside a component of a bipartite Ferrers-graph, all $\bigwedge$-irreducibles are connected to other ones or bound to the outside. This enriches the induced relation of that component in a satisfactory way. It becomes "rigid" enough to be transitive, i.e. a strict order.

Proposition 6.20 *Let $\mathbb{K} = (J, M, \leq)$ be the standard context of a lattice $\mathfrak{V}$ and Γ its Ferrers-graph with a bipartition $\mathfrak{L} \cup \mathfrak{R}$. Let $\Gamma_\circ$ be a component of Γ and $\tilde{L}_\circ$ its induced relation (w.r.t. $\mathfrak{L}$ and $\mathfrak{R}$). Then $\tilde{L}_\circ$ is a strict order.*

Proof: The relation $\tilde{L}_\circ$ is asymmetric since $\tilde{L}$ is (see Lemma 6.4).

Let $m_1, m_2, m_3 \in M$. We have to show $m_1 \, \tilde{L}_\circ \, m_2 \, \tilde{L}_\circ \, m_3 \implies m_1 \, \tilde{L}_\circ \, m_3$:

Since we know $\Gamma_\circ = \Gamma_{m_1, m_2} = \Gamma_{m_2, m_3}$ we notice by applying Lemma 6.18 that (m_1, m_2, m_3) is not a free triple. With Lemma 6.4 we find $m_1 \parallel m_3$.

If m_1, m_2 or m_2, m_3 are connected in $[\underline{v}, \overline{v}] = [m_1 \wedge m_2 \wedge m_3, m_1 \vee m_2 \vee m_3]$ then we find $m_1 \, \tilde{L}_\circ \, m_3$ by applying Lemma 6.15. The case that only m_1, m_3 are connected in $[\underline{v}, \overline{v}]$ can not occur: Then we had $[\underline{v}, \overline{v}] = [m_1 \wedge m_2, m_1 \vee m_2]$ and concluded with Lemma 6.14 that $m_1 \, \tilde{L}_\circ \, m_2 \iff m_3 \, \tilde{L}_\circ \, m_2$ contradicting our precondition.

If otherwise m_1 or m_3 is bound then we apply Lemma 6.16 and notice $m_1 \, \tilde{L}_\circ \, m_3$. If m_2 is bound we conclude $m_1 \, \tilde{L}_\circ \, m_2 \iff m_3 \, \tilde{L}_\circ \, m_2$, again a contradiction. We conclude in all possible cases $m_1 \, \tilde{L}_\circ \, m_3$. Therefore $\tilde{L}_\circ$ is transitive, i.e. a strict order. □

Proposition 6.20 enables us to search for left-orders on lattices possessing a bipartite Ferrers-graph Γ: Since the components of Γ provide parts of the induced relation that are strict orders, we can hope to find a strategy to flip over the components until we find a suitable configuration. We will come back to this issue in Section 6.3.

6.2.3 Edge Sequences in Γ and Connections in $\mathfrak{V}$

In this subsection we will highlight the coherence between connections in a lattice $\mathfrak{V}$ and edge sequences in its corresponding Ferrers-graph Γ in detail. It will turn out that m-components (see Definition 6.12) are an adequate instrument to describe the set of $\bigwedge$-irreducibles $M(\Gamma_\circ)$ contained in a component $\Gamma_\circ$ of Γ.

Our first result embarks the easy way from the Ferrers-graph to the lattice: Edge sequences of Γ represent connections in $\mathfrak{V}$ in the interval $I(\Gamma)$.

Lemma 6.21 *[Zsc08] Let $\mathfrak{V}$ be a lattice with Ferrers-graph Γ. Let*

$$(j_0, n_0)E(j_1, n_1)E \ldots E(j_{r-1}, n_{r-1})E(j_r, n_r)$$

be an edge sequence in a component $\Gamma_\circ$ of the Ferrers-graph. Then n_0 is connected to either n_{r-1} or n_r in the interval $[\underline{v}, \overline{v}]$ of $\Gamma_\circ$.

6.2 The Components of Ferrers-graphs

Proof: By Definition 6.9 we note that $p := n_0 \geq j_1 \leq n_2 \geq j_3 \leq n_4 \ldots \leq n_s$ is a connection in $\mathfrak{V}$ with either $s = r-1$ (if r is odd) or $s = r$ (if r is even). Let $[\underline{v}, \overline{v}]$ be the interval of $\Gamma_\circ$. With Lemma 6.8 we know that $j_i > \underline{v}$ and $n_i < \overline{v}$ holds for all $i \in \{0, \ldots, r\}$. Hence p is a connection of n_0 and n_s in $[\underline{v}, \overline{v}]$. □

See the picture on the right for an example: The edge sequence

$$(j_0, n_0) E(j_1, n_1) E(j_2, n_2) E(j_3, n_3) E(j_4, n_4)$$

is represented by the graph in the diagonal and the connection $p = n_0, j_1, n_2, j_3, n_4$ by the thick crosses. More precisely, every edge sequences determines two connections; the other one given by the thin crosses connects n_1 and n_3 via j_2.

	n_0	n_1	n_2	n_3	n_4
j_0	○	×			
j_1	×	○	×		
j_2		×	○	×	
j_3			×	○	×
j_4				×	○

The other way round can not proceeded in such a straightforward manner. An edge sequence in Γ can be constructed by two connections in an interval only. Additionally the elements of the connections have to meet further conditions. However, finding those edge sequences will enable us to describe the components of the Ferrers-graph by underlying lattice structures.

The following first result can intuitively be understood quite simple in a cross table of a standard context: If we find in a component $\Gamma_\circ$ a vertex in the column corresponding to m that is connected to an edge between two vertices situated in the columns corresponding to m_1 and m_2 then there exists also an edge in $\Gamma_\circ$ between vertices in the columns of m and m_1 or m and m_2.

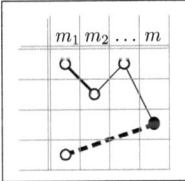

Lemma 6.22 *[Zsc08] Let $\mathbb{K} = (J, M, \leq)$ be the standard context of a lattice $\mathfrak{V}$ with Ferrers-graph Γ. Furthermore let $m, m_1, m_2 \in M$, s.t. $m \in M(\Gamma_{m_1, m_2})$ and m is incomparable to m_1 or m_2. Then $\Gamma_{m_1, m_2} = \Gamma_{m, m_1}$ or $\Gamma_{m_1, m_2} = \Gamma_{m, m_2}$.*

Proof: If there is a connection between m and m_1 or between m and m_2 in the interval $I := [m_1 \wedge m_2, m_1 \vee m_2]$ then we may apply Lemma 6.14 to prove the claim.

Since $m \in M(\Gamma_{m_1, m_2})$, we find for adequate $\bigvee$-irreducibles $h_i \in J$ and $\bigwedge$-irreducibles $n_i \in M$, (for all indices $i \in \{1, \ldots, r\}$ an edge sequence

$$K := (h_1, n_1) E (h_2, n_2) E \ldots E (h_{r-1}, n_{r-1}) E (h_r, n_r)$$

with $n_1 = m_1$, $n_2 = m_2$ and $n_r = m$. Clearly $\Gamma_{n_1, n_2} = \Gamma_{n_{r-1} n_r}$. W.l.o.g. (in case of $2 \nmid r$)

$$p = n_1 \geq h_2 \leq n_3 \geq \ldots \geq h_{r-1} \leq n_r \text{ and}$$
$$q = n_2 \geq h_1 \leq n_2 \geq h_3 \leq \ldots$$
$$\ldots \leq n_{r-1} \geq h_r \leq n_{r-1}$$

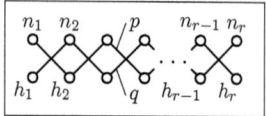

are connections in $I(\Gamma_{m_1,m_2})$ (see Lemma 6.21). Consider the sequence

$$S = (h_r, n_r), (h_{r-1}, n_{r-1}), (h_{r-2}, n_r), (h_{r-1}, n_{r-3}), (h_{r-4}, n_r),$$
$$\ldots, (h_{r-1}, n_2), (h_1, n_r).$$

1. If the conditions $h_i \parallel n_r$ and $h_{r-1} \parallel n_j$ hold for all $i \in \{1, \ldots, r\} \setminus \{r-1\}$ and $j \in \{1, \ldots, r\} \setminus \{r, r-2\}$ then S is an edge sequence in Γ. We conclude $\Gamma_{n_1,n_2} = \Gamma_{n_{r-1},n_r} = \Gamma_{n_2,n_r}$.

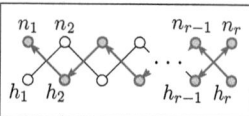

2. Otherwise let t be the smallest index satisfying $h_t \leq n_r$ or $h_{r-1} \leq n_t$ (recall that $h_t > n_r \implies h_{r-1} \leq n_{t\pm 1}$ and $h_{r-1} > n_t \implies h_{t\pm 1} \leq n_r$).

   (a) Let $h_t \leq n_r$: In case of $t \leq 2$ we find the connection $m_{3-t} \geq h_t \leq m$ in I. Otherwise $t > 2$ and

   $$(h_t, n_t)E(h_{t-1}, n_r)E(h_{r-1}, n_{t-2})E(h_{t-3}, n_r)E \ldots E(h_y, n_r)E(h_{r-1}, n_x)$$

   with $y = (t \pmod 2) + 1$ and $x = 3 - y$ is an edge sequence in $\Gamma_{n_y,n_2} = \Gamma_{n_1,n_2}$ since $(h_t, n_t) \in V(\Gamma_{n_1,n_2})$ (see picture; $y = 1$).

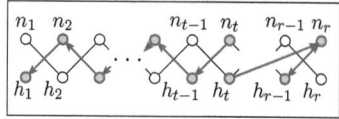

   (b) Let $h_{r-1} \leq n_t$:

   (b1) $t > 2$: Then

   $$(h_t, n_t)E(h_{r-1}, n_{t-1})E(h_{t-2}, n_r)E(h_{r-1}, n_{t-3})E \ldots$$
   $$\ldots E(h_{r-1}, n_y)E(h_x, n_r)$$

   with $y = (t \pmod 2) + 1$ and $x = 3 - y$ is an edge sequence in $\Gamma_{n_y,n_2} = \Gamma_{n_1,n_2}$ since $(h_t, n_t) \in V(\Gamma_{n_1,n_2})$ (see picture; $y = 1$).

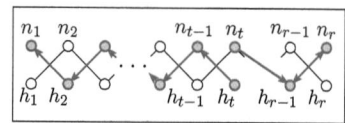

6.2 The Components of Ferrers-graphs

(b2) $t \in \{1, 2\}$. We consider only the case $h_{r-1} \leq n_2 \iff h_{r-1} \leq n_1$, otherwise we had a connection $n_t \geq h_{r-1} \leq n_r$ in I proving our claim. By precondition we know $n_1 \parallel n_r$ or $n_2 \parallel n_r$. W.l.o.g. let $n_1 \parallel n_r$, then we find a $\bigvee$-irreducible j satisfying $n_r \geq j \parallel n_1$.
If $j \not\leq n_{r-1}$ then we extend K by two nodes and gain the sequence $\tilde{K} = (h_1, n_1)E \ldots E(h_r, n_r)E(j, n_{r-1})E(h_r, n_r)$. For the appropriate sequence

$$\tilde{S} = (h_r, n_r), (j, n_{r-1}), (h_{r-2}, n_r), (j, n_{r-3}), (h_{r-4}, n_r),$$
$$\ldots, (j, n_2), (h_1, n_r)$$

we can apply one of the other cases 1. or 2.(a) or 2.(b1).
Let otherwise $j \leq n_{r-1}$. We again have to distinguish two cases.

First, if $h_2 \leq n_{r-1}$ then we find an edge sequence $(j, n_1)E(h_2, n_r)E(h_{r-1}, n_{r-1})$ (or a connection $n_1 \geq h_2 \leq n_r$ in I). We conclude $\Gamma_{n_1, n_r} = \Gamma_{n_r, n_{r-1}} = \Gamma_{n_1, n_2}$.

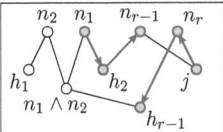

Finally assume $h_2 \not\leq n_{r-1}$, then $(h_2, n_r)E(j, n_1)E(h_2, n_{r-1})$ is an edge sequence, i.e. $\Gamma_{n_1, n_r} = \Gamma_{n_1, n_{r-1}}$. Moreover, by the construction of K, also n_{r-1} is an element of $M(\Gamma_{n_1, n_2})$. We observe that $n_1 \geq h_{r-1} \leq n_r$ is a connection in $[n_r \wedge n_{r-1}, n_r \vee n_{r-1}]$.

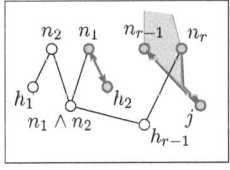

With Lemma 6.14 we conclude $\Gamma_{n_r, n_{r-1}} = \Gamma_{n_1, n_r}$ or $\Gamma_{n_r, n_{r-1}} = \Gamma_{n_1, n_{r-1}}$ which both implies $\Gamma_{n_1, n_2} = \Gamma_{n_1, n_r}$.

□

In the next lemma, we will find a first coherence between m-components of a lattice $\mathfrak{V}$ and components of its respective Ferrers-graph Γ. More precisely, all $\bigwedge$-irreducibles that are connected in the interval of a component $\Gamma_\circ$ to an $\bigwedge$-irreducible in $\Gamma_\circ$ (i.e. an element of $M(\Gamma_\circ)$) are itself contained in $M(\Gamma_\circ)$.

Lemma 6.23 *[Zsc08] Let $\mathbb{K} = (J, M, \leq)$ be the standard context of a lattice $\mathfrak{V}$ and Γ its Ferrers-graph. Let $m_1, m_2 \in M$, s.t. Γ_{m_1, m_2} is a component of Γ. Then $U_{m_1}(I(\Gamma_{m_1, m_2})) \subseteq M(\Gamma_{m_1, m_2})$.*

Proof: Let $[\underline{v}, \overline{v}] := I(\Gamma_{m_1, m_2})$. Furthermore let m be a $\bigwedge$-irreducible meeting $m \in U_{m_1}([\underline{v}, \overline{v}])$. We show $m \in M(\Gamma_{m_1, m_2})$:

94   **Chapter 6: Left-relations on Contexts**

If m and m_1 are also connected in $[m_1 \wedge m_2, m_1 \vee m_2]$ then by Lemma 6.14 we may conclude w.l.o.g. $\Gamma_{m_1,m_2} = \Gamma_{m,m_2}$. This implies $m \in M(\Gamma_{m_1,m_2})$ (see Definition 6.7).

Let otherwise $q = m \geq h_1 \leq n_1 \geq \ldots \leq n_{r-1} \geq h_r \leq n_r =: m_1$ be a shortest connection (for adequately chosen $h_i \in J$, $n_i \in M$, $i \in \{1, \ldots, r\}$) between m and m_1 in $I(\Gamma_{m_1,m_2})$ containing a $\bigvee$-irreducible $h_k \leq m_{12} := m_1 \wedge m_2$ (then $k = r$ since q is a shortest connection) or a $\bigwedge$-irreducible $n_k \geq m^{12} := m_1 \vee m_2$ (then $k = r-1$). We may exclude the case that both inequalities hold since then we could find a $\bigvee$-irreducible $\tilde{j}$ with $n_{r-1} \geq \tilde{j} \leq n_r$ and $\tilde{j} > m_{12}$. We consider the case $h_r \leq m_{12}$, $n_{r-1} \not\geq m^{12}$ and omit the dual $h_r \not\leq m_{12}$, $n_{r-1} \geq m^{12}$. Recall that

$$m_2 \parallel h_k \parallel m_1 \text{ holds for all } k \in \{1, \ldots, r-1\} \tag{6.2}$$

since we had a connection $\tilde{q} = m \geq h_1 \leq \ldots \geq h_k \leq m_1(m_2)$ in $[m_{12}, m^{12}]$ otherwise. Similarly

$$m_2 \parallel n_k \parallel m_1 \text{ holds for all } k \in \{1, \ldots, r-1\}. \tag{6.3}$$

Since q is a connection in $[\underline{v}, \overline{v}]$, we note $h_r \not\leq \underline{v}$, i.e. $n_{r-1} \wedge m_{12} \in I(\Gamma_{m_1,m_2})$. According to Definition 6.7 there exists a $\bigwedge$-irreducible $\tilde{m} \in M(\Gamma_{m_1,m_2})$ with $n_{r-1} \wedge m_{12} \parallel \tilde{m}$. Hence, there exists an edge sequence (with $m_t := \tilde{m}$)

$$(j_1, m_1)E(j_2, m_2)E \ldots E(j_{t-1}, m_{t-1})E(j_t, m_t)$$

for suitably chosen $\bigvee$-irreducibles j_i and $\bigwedge$-irreducibles m_i ($i \in \{1, \ldots, t\}$).

1. If there exists an index $i \in \{3, \ldots, t\}$ with $n_{r-1} \wedge m_{12} \parallel m_i \parallel m_{12}$ then we note w.l.o.g. $\Gamma_{m_1,m_2} = \Gamma_{m_i,m_2}$ due to Lemma 6.22. Also, we find a $\bigvee$-irreducible $\tilde{j}$ with $m_i \parallel \tilde{j} \leq n_{r-1} \wedge m_{12}$. Then, $\tilde{q} = m \geq h_1 \leq \ldots \leq n_{r-1} \geq \tilde{j} \leq m_2$ is a connection in $[m_i \wedge m_2, m_i \vee m_2]$:
   $n_k \geq m_2 \vee m_i$ contradicts Condition 6.3,
   $h_k \leq m_2 \wedge m_i$ contradicts Condition 6.2 and
   $\tilde{j} \leq m_2 \wedge m_i$ is not possible since we chose
   $\tilde{j} \parallel m_i$. Hence, $m \in M(\Gamma_{m_i,m_2}) = M(\Gamma_{m_1,m_2})$.

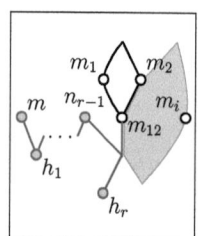

6.2 The Components of Ferrers-graphs

2. Otherwise we find an index $i \in \{2, \ldots, t-1\}$ with $m_i \geq n_{r-1} \wedge m_{12} \parallel m_{i+1}$ (recall $m_i \parallel m_{i+1}$ and $m_1 \geq n_{r-1} \wedge m_{12} \parallel m_t$. Again, we find a $\vee$-irreducible $\tilde{j}$ with $m_{i+1} \parallel \tilde{j} \leq n_{r-1} \wedge m_{12}$. Then $\tilde{q} = m \geq h_1 \leq \ldots \leq n_{r-1} \geq \tilde{j} \leq m_2$ is a connection in $[m_i \wedge m_{i+1}, m_i \vee m_{i+1}]$:
First, $n_k \geq m_i \vee m_{i+1} \implies n_k \geq m_i \geq h_r$, i.e. $k = r-1$ contradicts $m_{12} \geq m_i \geq n_{r-1} \wedge m_{12}$.
Second, $h_k \leq m_i \wedge m_{i+1} \implies h_k \leq m_{12}$ contradicts q to be a shortest connection and $\tilde{j} \leq m_i \wedge m_{i+1}$ is not possible since we assumed $\tilde{j} \parallel m_{i+1}$. Hence, $m \in M(\Gamma_{m_i, m_{i+1}}) = M(\Gamma_{m_1, m_2})$. □

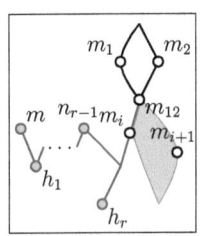

Now, we are prepared for the main result of this subsection. Additionally to Lemma 6.23, all $\wedge$-irreducibles of a component Γ_o of a Ferrers-graph Γ are connected in the interval of Γ_o. Thus, we can describe components of Γ quite comfortably by m-components. More generally, the following proposition allows to describ graph properties by lattice properties.

Proposition 6.24 *[Zsc08] Let $\mathbb{K} = (J, M, \leq)$ be the standard context of a lattice with Ferrers-graph Γ. Let $m_1 \parallel m_2 \in M$. Then*

$$M(\Gamma_{m_1, m_2}) = U_{m_1}(I(\Gamma_{m_1, m_2})) \cup U_{m_2}(I(\Gamma_{m_1, m_2})).$$

Proof: Lemma 6.23 states $M(\Gamma_{m_1, m_2}) \supseteq U_{m_1}(I) \cup U_{m_2}(I)$.
Let $m \in M(\Gamma_{m_1, m_2})$. Then there exists an edge sequence

$$(j, m) E \ldots E(j_1, m_1) E(j_2, m_2)$$

for adequate $j, j_1, j_2 \in J$. According to Lemma 6.21, m is connected to m_1 or m_2, i.e. $m \in U_{m_1}(I) \cup U_{m_2}(I)$. Hence $M(\Gamma_{m_1, m_2}) \subseteq U_{m_1}(I) \cup U_{m_2}(I)$. □

Finally, we want to verify Proposition 6.24 in an example. Consider the lattice given by a diagram on the right. We easily check that we find two intervals providing m-components, namely $[0, 1]$ and $[e, 1]$ with

$$U_a([0,1]) = \{a\}, U_b([0,1]) = \{b,c\}, U_d([0,1]) = \{d\}$$
$$U_b([e,1]) = \{b\}, U_c([e,1]) = \{c\}.$$

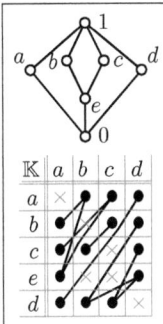

In the standard context, we find four components: Γ_{ab}, Γ_{ad} and Γ_{bd} have the interval $[0,1]$ and Γ_{bc} has $[e,1]$. Due to Proposition 6.24, $M(\Gamma_{ab}) = \{a,b,c\}, M(\Gamma_{ad}) = \{a,d\}, M(\Gamma_{bd}) = \{b,c,d\}$ and $M(\Gamma_{bc}) = \{b,c\}$.

6.2.4 Components of Type 1 and Type 2

In this subsection, we introduce a distinction of the components of a Ferrers-graph into two types. This will enable us to investigate coherences to m-components on the respective lattice more precisely than in Proposition 6.24. Eventually, we need this preparations in order to count the number of left-orders on a lattice in Section 6.3.2. We will obtain these left-orders by turning around components of the Ferrers-graph in the "right manner".

In the following lemma, we make an observation that is a consequence of Lemma 6.21. Sloppily spoken, edge sequences of the Ferrers-graph cause connections in the respective lattice.

Lemma 6.25 *[Zsc08] Let $\mathbb{K} = (J, M, \leq)$ be the standard context of a lattice $\mathfrak{V}$ with the Ferrers-graph Γ. Let $m_1, m_2, m_3, m_4 \in M$, s.t. $\Gamma_{m_1,m_2} = \Gamma_{m_3,m_4}$ is a component of Γ possessing the interval $[\underline{v}, \overline{v}]$. Then m_1, m_2 are connected in $[\underline{v}, \overline{v}]$ if and only if m_3, m_4 are connected in $[\underline{v}, \overline{v}]$.*

Proof: Since $\Gamma_{m_1,m_2} = \Gamma_{m_3,m_4}$, there exists an edge sequence

$$(h_1, m_1) E(h_2, m_2) E \ldots E(h_3, m_3) E(h_4, m_4)$$

By Lemma 6.21 this refers w.l.o.g. to a connection between m_1 and m_3 and m_2 and m_4 in $[\underline{v}, \overline{v}]$ respectively. If m_1 and m_2 are connected, we find a connection

$$p := m_3 \ldots m_1 \ldots m_2 \ldots m_4$$

in $[\underline{v}, \overline{v}]$. Since $m_3, m_4 \in M(\Gamma_{m_3,m_4})$, both $\bigwedge$-irreducibles are connected. Dually, if m_3, m_4 are connected, then m_1, m_2 are, too. □

Now we are able to introduce the already mentioned partition. The reason for the name we chose for the elements of the two classes will be discovered in Corollary 6.27.

Definition 6.26 *[Zsc08] Let Γ be the Ferrers-graph of a lattice $\mathfrak{V}$. A component $\Gamma_{m,n}$ is of type 1 if m and n are connected in $I(\Gamma_{m,n})$. Otherwise $\Gamma_{m,n}$ is of type 2.* ◊

Definition 6.26 is well-defined, in particular a component is either of type 1 or of type 2. This is a direct consequence of Lemma 6.25.

See Figure 6.10 for an example of different types of components. In the standard context one can encounter two components $\Gamma_{n_1,n_2} = \Gamma_{n_1,n_3} = \Gamma_{n_1,n_4}$ and $\Gamma_{n_2,n_3} = \Gamma_{n_3,n_4}$. Both possess the interval $[0_{\mathfrak{V}}, 1_{\mathfrak{V}}]$. We observe that Γ_{n_1,n_2} is of type 2 since m_1 is not connected to one of the other $\bigwedge$-irreducible elements.

6.2 The Components of Ferrers-graphs

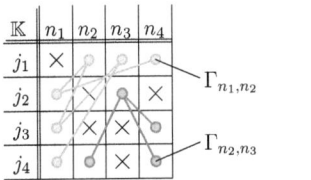
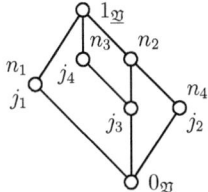

Figure 6.10: The Ferrers-graph of the standard context of this lattice consists of two components. Thereby Γ_{n_1,n_2} is of type 2 and Γ_{n_2,n_3} of type 1.

In contrast, we notice that n_2 and n_3 are connected in $[0_{\mathfrak{V}}, 1_{\mathfrak{V}}]$ via j_3. Hence, Γ_{n_2,n_3} is of type 1. Also n_2, n_4 (via j_2) and n_3, n_4 (via j_3, n_2, j_2) are connected in that interval.

Combining Definition 6.26 with the characterization of components of the Ferrers-graph by appropriate m-components in Proposition 6.24 will explain the notations "type 1" and "type 2" we chose. We will see that the set of $\bigwedge$-irreducibles $M(\Gamma_\circ)$ of a type 1 component is equal to 1 such m-component. If otherwise $\Gamma_\circ$ is of type 2, then $M(\Gamma_\circ)$ equals the (disjoint) union of 2 of them.

Corollary 6.27 *[Zsc08] Let $\mathfrak{V}$ be a lattice with Ferrers-graph Γ. Let Γ_{m_1,m_2} be a component of Γ possessing the interval $I := I(\Gamma_{m_1,m_2})$.*

1. *If Γ_{m_1,m_2} is of type 1 then $M(\Gamma_{m_1,m_2}) = U_{m_1}(I) = U_{m_2}(I)$.*

2. *If Γ_{m_1,m_2} is of type 2 then $M(\Gamma_{m_1,m_2}) = U_{m_1}(I) \;\dot\cup\; U_{m_2}(I)$.*

Proof: By Definition 6.26 we immediately see $U_{m_1}(I) = U_{m_2}(I)$ for type 1 and $U_{m_1}(I) \cap U_{m_2}(I) = \emptyset$ for type 2. The claim follows with Proposition 6.24. □

See Figure 6.11 for a schematic depiction of components of type 1 (on the left) and type 2 (on the right) in the lattice perspective. The sets $J(\Gamma_\circ)$ and $M(\Gamma_\circ)$ of a type 1 component $\Gamma_\circ$ are all situated in *one* proper component of the lattice $\mathfrak{V}$. Contrary, in a type 2 component these sets are spread on *two* proper components.

Components of type 1 and of type 2 can be easily distinguished when calculating their intervals. Thereby a type 2 component $\Gamma_{m,n}$ is the less complicated candidate, instead of determining $I(\Gamma_{m,n})$ by $[\bigwedge M(\Gamma_{m,n}), \bigvee M(\Gamma_{m,n})]$ according to Definition 6.6 it is enough to consider $[m \wedge n, m \vee n]$ only, as the following lemma states.

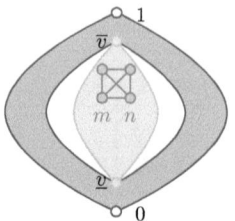 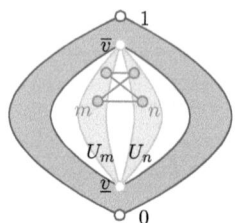

Figure 6.11: How to find components of type 1 and type 2 in the underlying lattice.

In fact, also in type 1 components the limits of $I(\Gamma_{m,n})$ can be determined as intersection and union of only two elements of $M(\Gamma_{m,n})$. However, these elements can not be arbitrarily chosen.

Lemma 6.28 *[Zsc08] Let $\Gamma_{m,n}$ be a component of the Ferrers-graph of a lattice and $I(\Gamma_{m,n}) = [\underline{v}, \overline{v}]$.*

1. *If $\Gamma_{m,n}$ is of type 2 then $[m \wedge n, m \vee n] = [\underline{v}, \overline{v}]$.*

2. *If $\Gamma_{m,n}$ is of type 1 then there exists an edge $\{(j_1, m_1), (j_2, m_2)\}$ in $\Gamma_{m,n}$ with $m_1 \wedge m_2 > \underline{v}$.*

Proof:

1. Let $m \wedge n > \underline{v}$. Then we find a $\bigvee$-irreducible j satisfying $j \leq m \wedge n$ and $j \not\leq \underline{v}$. Hence m and n are connected in $[\underline{v}, \overline{v}]$, i.e. $\Gamma_{m,n}$ is not of type 2. In analogy, if $m \vee n < \overline{v}$ then there exists a $\bigwedge$-irreducible $\tilde{n}$ satisfying $\tilde{n} \geq m \vee n$ and $\tilde{n} \not\geq \overline{v}$. Hence m and n are connected via $p = m \geq j \leq \tilde{n} \geq h \leq n$ in $[\underline{v}, \overline{v}]$.

2. The second fact is shown by induction over the length of a shortest connection p between m and n in $[\underline{v}, \overline{v}]$.

   If p is of length one, i.e. $p = m \geq j \leq n$ then $m \wedge n \geq j \not\leq \underline{v}$, hence $m \wedge n > \underline{v}$.

   If p is of length $r > 1$, i.e. $p = n_0 \geq h_1 \leq n_1 \geq \ldots \geq h_r \leq n_r$ (with $m = n_0$ and $n = n_r$) then n_1 is connected to m in $[m \wedge n, m \vee n]$. We conclude with Lemma 6.14 that $\Gamma_{m,n} = \Gamma_{m,n_1}$ or $\Gamma_{m,n} = \Gamma_{n,n_1}$. In both cases the connections between m and n_1 and n and n_1 respectively are shorter than p. Hence we find a pair of $\bigwedge$-irreducibles meeting the requirements of the claim. □

6.2 The Components of Ferrers-graphs

Let us further consider the the interval $I(\Gamma_{m,n})$ of a component $\Gamma_{m,n}$ of a Ferrers-graph. By Definition 6.6 we know that $[m \wedge n, m \vee n]$ is a subset of $I(\Gamma_{m,n})$. Now we want to find out, when we may flip over the inclusion sign.

Lemma 6.29 *Let $\mathbb{K} = (J, M, \leq)$ be the standard context of a lattice $\mathfrak{V}$ possessing the Ferrers-graph Γ. Let $m \parallel n \in M$ be both non-bound in $[\underline{v}, \overline{v}] := [m \wedge n, m \vee n]$. Then $I(\Gamma_{m,n}) \subseteq [\underline{v}, \overline{v}]$.*

Proof: Let $\tilde{m} \in M(\Gamma_{m,n})$. Hence there exists (for adequate chosen $\{h_i \mid i \in \{1, \ldots, r\} \subseteq J$ and $\{n_i \mid i \in \{1, \ldots, r\} \subseteq M$) an edge sequence

$$(h_1, n_1)(h_2, n_2) \ldots (h_r, n_r)$$

in $\Gamma_{m,n}$ with $n_1 = m$, $n_2 = n$ and $n_r = \tilde{m}$. According to Remark 6.11 and Lemma 6.21, for all $k \in \{1, \ldots, r\}$ the $\bigwedge$-irreducible n_k is non-bound.

Suppose that, for an index $k \in \{2, \ldots, r-1\}$, both $n_{k-1} > \underline{v}$ and $n_k > \underline{v}$ hold. Then $j_k \parallel \underline{v}$ contradicts n_{k-1} to be non-bound and $j_k \leq \underline{v}$ contradicts $j_k \parallel n_k$. Hence, $n_{k+1} \geq j_k > \underline{v}$. Since $n_1, n_2 > \underline{v}$, we conclude $\tilde{m} > \underline{v}$. In analogy we may realize that $\tilde{m} < \overline{v}$. Hence, $M(\Gamma_{m,n}) \subseteq [\underline{v}, \overline{v}]$. The claim follows with Definition 6.6. $\square$

We have seen in Corollary 6.27 that every type 2 component $\Gamma_{m,n}$ of a Ferrers-graph Γ determines two m-components. The following lemma states that the converse holds as well: every pair of m-components over the same interval uniquely determines a type 2 component of Γ:

Lemma 6.30 *Let $\mathbb{K} = (J, M, \leq)$ be the standard context of a lattice $\mathfrak{V}$ with the Ferrers-graph Γ. Let $[\underline{v}, \overline{v}]$ be an interval in $\mathfrak{V}$ and $m, n \in M$, satisfying $U_m([\underline{v}, \overline{v}]) \neq U_n([\underline{v}, \overline{v}])$. Then $\Gamma_{m,n}$ is of type 2 and $I(\Gamma_{m,n}) = [\underline{v}, \overline{v}]$.*

In particular, if (m, n, o) is a free triple then $\Gamma_{m,n}$ is of type 2 and $I(\Gamma_{m,n}) = [m \wedge n \wedge o, m \vee n \vee o]$.

Proof: Due to Definition 6.12 we know $m, n \in [\underline{v}, \overline{v}]$ i.e., $[m \wedge n, m \vee n] \subseteq [\underline{v}, \overline{v}]$. With Remark 6.11 we conclude $[m \wedge n, m \vee n] = [\underline{v}, \overline{v}]$ and with Definition 6.7 $I(\Gamma_{m,n}) \supseteq [\underline{v}, \overline{v}]$.

By Definition 6.12 m and n are non-bound in $[\underline{v}, \overline{v}]$. By applying Lemma 6.29 we find $I(\Gamma_{m,n}) \subseteq [\underline{v}, \overline{v}]$. Hence, $I(\Gamma_{m,n}) = [\underline{v}, \overline{v}]$ and with Definition 6.26 we conclude that $\Gamma_{m,n}$ is of type 2.

If (m, n, o) is a free triple then $U_m([m \wedge n \wedge o, m \vee n \vee o]) \neq U_n([m \wedge n \wedge o, m \vee n \vee o])$. The claim follows with the first statement. $\square$

In Figure 6.12 we show the differences between type 1 and type 2 components in the context perspective. We are given a lattice on the left together with its standard context $\mathbb{K} = (J, M, \leq)$. Its Ferrers-graph consists of two components, thereby $\Gamma_o = \Gamma_{m,n}$ is of type 2 and $\Gamma_1 = \Gamma_{n,o}$ is of type 1. In the context, a type 2 component has edges between columns of different m-components only. In case of Γ_0, the involved m-components are $U_m([0_\mathfrak{M}, 1_\mathfrak{M}]) = \{m\}$ and $U_n([0_\mathfrak{M}, 1_\mathfrak{M}]) = \{n, o, p, q\}$. Hence, we find in $\Gamma_{m,n}$ only edges between the first column and the other columns. In Γ_1, every pair of columns belonging to incomparable $\bigwedge$-irreducibles is connected by an edge.

We may reveal this difference even better by introducing a graph $\hat{\Gamma}_o$ corresponding to a component Γ_o of Γ in the following way: The vertices of $\hat{\Gamma}_o$ are the columns of $\mathbb{K}$ that are touched by Γ_o. Two vertices of $\hat{\Gamma}_o$ are connected by an edge, if the respective columns are connected by an edge of Γ_o. Formally, we define:

$$V(\hat{\Gamma}_o) := M(\Gamma_o) \text{ and}$$
$$E(\hat{\Gamma}_o) := \{(m, n) \in M^2 \mid \exists j, h \in J : \{(j, m), (h, n)\} \in E(\Gamma_o)\}.$$

We realize that the graph $\hat{\Gamma}_0$ is bipartite with the vertex classes $U_m([0_\mathfrak{M}, 1_\mathfrak{M}])$ and $U_n([0_\mathfrak{M}, 1_\mathfrak{M}])$. Meanwhile, two vertices in $\hat{\Gamma}_1$ are connected if and only if the appropriate $\bigwedge$-irreducibles are incomparable, i.e. $E(\hat{\Gamma}_1) = \|_{V(\Gamma_1)}$.

At least the first observation always holds; we may subsume to the following

Remark 6.31 *If $\Gamma_{m,n}$ is a component of type 2 then the graph $\hat{\Gamma}_{m,n}$ is bipartite with the vertex classes $U_m(I(\Gamma_{m,n}))$ and $U_m(I(\Gamma_{m,n}))$.*

If $\Gamma_{m,n}$ is a component of type 1 then $E(\hat{\Gamma}_{m,n}) \subseteq \|_{V(\Gamma_{m,n})}$.

The first fact follows from Lemma 6.28 and Lemma 6.14. The second fact is trivial since it is a direct consequence of Definition 6.6. However, it can not be formulated in a stricter form similar to the first.

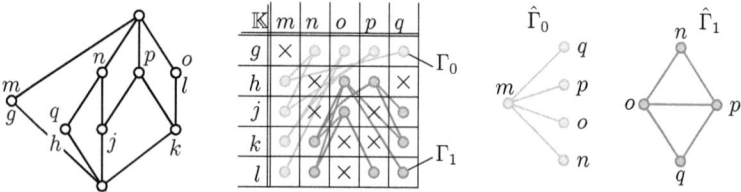

Figure 6.12: A lattice together with its standard context $\mathbb{K}$. The Ferrers-graph consists of the components Γ_0 of type 2 and Γ_1 of type 1. On the right the respective graphs $\hat{\Gamma}_0$ and $\hat{\Gamma}_1$ are depicted.

6.2 The Components of Ferrers-graphs

We found out already that for a type 2 component Γ_o, the set $M(\Gamma_o)$ of its contained $\bigwedge$-irreducibles is the disjoint union of two m-components U_m and U_n. We can describe the structure of Γ_o even more accurate; every pair $(m_1, n_1) \in U_m \times U_n$ is connected in Γ_o, i.e. there exists an edge between them in the graph $\hat{\Gamma}_o$, as we see now:

Lemma 6.32 *[Zsc07a] Let $\mathbb{K} = (J, M, \leq)$ be the standard context of a lattice $\mathfrak{V}$ with the Ferrers-graph Γ. Let $m \parallel n \in M$, s.t. $\Gamma_{m,n}$ is of type 2. Then*

$$\Gamma_{m,n} = \Gamma_{m_1, n_1} \iff m_1 \in U_m(I(\Gamma_{m,n})) \text{ and } n_1 \in U_n(I(\Gamma_{m,n})) \text{ (or vice versa).}$$

Furthermore, if Γ is bipartite and one of the statements above is true then $m \, \tilde{L} \, n \iff m_1 \, \tilde{L} \, n_1$ holds for the respective induced relation $\tilde{L}$.

Proof: "$\Rightarrow$": Let $\Gamma_{m,n} = \Gamma_{m_1, n_1}$. Hence there exists an edge sequence

$$(g_1, m_1) E (h_1, n_1) E \ldots E(g, m) E(h, n).$$

By Definition 6.7 we observe $m_1, n_1 \in M(\Gamma_{m,n})$. Applying Lemma 6.21 w.l.o.g. $m_1 \in U_m(I(\Gamma_{m,n}))$ and $n_1 \in U_n(I(\Gamma_{m,n}))$. If Γ is additionally bipartite then

$$m \, \tilde{L}_j \, n \Longrightarrow g \, \mathfrak{R}_j \, m \Longrightarrow \ldots \Longrightarrow h_1 \, \mathfrak{L}_j \, n_1 \Longrightarrow g_1 \, \mathfrak{R}_j \, m_1 \Longrightarrow m_1 \, \tilde{L}_j \, zn_1.$$

"$\Leftarrow$": By Lemma 6.28 we know $I(\Gamma_{m,n}) = [m \wedge n, m \vee n]$. Furthermore m_1 and n are not connected in $I(\Gamma_{m,n})$ since then also m and n would be connected (recall that connectedness in an interval is an equivalence relation). Hence we may apply Lemma 6.14 and conclude $\Gamma_{m,n} = \Gamma_{m_1,n}$. With an analog argumentation we find $\Gamma_{m_1,n} = \Gamma_{m_1, n_1}$. □

Combining Lemma 6.32 with Remark 6.31 we may conclude that the graph $\hat{\Gamma}_o$ is a complete bipartite graph if Γ_o is of type 2. In particular, if Γ_o is also bipartite with induced relation $\tilde{L}_o$ and interval I_o then we easily notice that $\hat{\Gamma}_o = (M(\Gamma_o), \tilde{L}_o \,\dot\cup\, \tilde{L}_o^{-1})$. For the associated directed graph $(M(\Gamma_o), \tilde{L}_o)$ we obtain $\tilde{L}_o = U_m(I_o) \times U_n(I_o)$ (see picture on the right; $\tilde{L}_o$ is symbolized by the arrows), or vice versa $\tilde{L}_o = U_n(I_o) \times U_m(I_o)$.

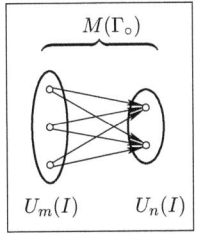

After collecting a lot of "little" results describing the structure of the components of a Ferrers-graph, we are now able to collect the fruits of our work.

6.3 Gaining Left-orders out of Ferrers-graphs

In the last section, we analyzed components of Ferrers-graphs. Thereby, we achieved two main results. First, the induced relation $\tilde{L}$ of each component $\Gamma_\circ$ can be extended to a strict order, and second, the $\bigwedge$-irreducibles of $\Gamma_\circ$ can be described comfortably by m-components of the underlying lattice.

Now we are interested in the behavior of $\tilde{L}$ if we turn around some components. As already mentioned, we mean by this phrase the process of changig the orientation of the induced relations $\tilde{L}_\circ$ of some components $\Gamma_\circ$. In subsection 6.3.1, we concern the question which of the components to flip over in order to make $\tilde{L}$ transitive, i.e. (see Lemma 6.5) the respective lattice planar.

Afterwards in Subsection 6.3.2, we consider a bipartition of Γ that induces already a left-order and try to find all possibilities to flip components while keeping the induced relation transisitve. Thus, we will be allowed to specify all non-similar plane diagrams of a lattice.

6.3.1 Turning Components of Γ

We start this subsection with an obvious observation: A bipartite graph G stays bipartite, if some of its components are "turned around"[2]. See Figure 6.13 for an intuition of this fact.

Remark 6.33 *Let $\Gamma = (V, E)$ be a bipartite graph with vertex classes X and Y and Γ_k, $k \in K$ its components. Let $X_k = X \cap V(\Gamma_k)$ and $Y_k = Y \cap V(\Gamma_k)$ be the vertex classes of the appropriate components Γ_k. Let $R_k \in \{X_k, Y_k\}$ for all $k \in K$. Then the sets $R = \bigcup_{k \in K} R_k$ and $V \setminus R$ are a bipartition of Γ.*

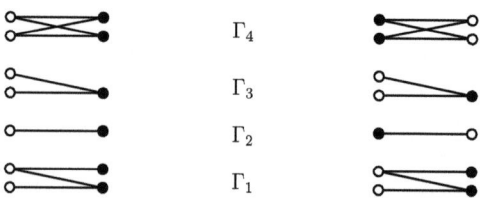

Figure 6.13: The bipartite graph Γ consists of four components. "Turning around" some of them (here Γ_2 and Γ_4) supplies a new bipartition.

---

[2] More general, the technique of "recoloring" vertices of graphs is used (without proof), for instance in *Brooks Theorem* [Die96].

6.3 Gaining Left-orders out of Ferrers-graphs

The following lemma gives a constructive method to obtain a left-order from a bipartite Ferrers-graph. Thereby, components are turned according to given linear orders on m-components sharing the same interval.

Lemma 6.34 *[Zsc07a] Let $\mathbb{K} = (J, M, \leq)$ be the standard context of a lattice $\mathfrak{V}$ and Γ its bipartite Ferrers-graph. Let Γ_k, $k \in K$ be the components of Γ and $\tilde{L}_k$ their respective induced relations. There exists a relation $\hat{L} = \biguplus_{k \in K} \hat{L}_k$ with $\hat{L}_k \in \{\tilde{L}_k, \tilde{L}_k^{-1}\}$ that is transitive.*

Proof: For each interval $[\underline{v}, \overline{v}] \subseteq \mathfrak{V}$ we introduce a linear order on its components:
$$U_{m_1} \leq U_{m_2} \leq \ldots \leq U_{m_t}. \tag{6.4}$$

Let $\Gamma_{m,n}$ be a component of type 2 with induced relation $\tilde{L}_{m,n}$. We define
$$\hat{L}_{m,n} := U_m(I) \times U_n(I).$$

According to Lemma 6.32 (and below) $\hat{L}_{m,n} \in \{\tilde{L}_{m,n}, \tilde{L}_{m,n}^{-1}\}$. For all relations $\tilde{L}_o$ induced by components Γ_o of type 1, we set $\hat{L}_o := \tilde{L}_o$.

Let $m \, \hat{L}_{m,n} \, n \, \hat{L}_{n,o} \, o$. By Lemma 6.4 we know $m \parallel o$. If $\hat{L}_{m,o}$ is an element of $\{\hat{L}_{m,n}, \hat{L}_{n,o}\}$ then we conclude with Lemma 6.18 and Corollary 6.17 that $m \, \hat{L}_{m,o} \, o$. Otherwise (m, n, o) is a free triple by Lemma 6.18.

Applying Lemma 6.30 we know that the components $\Gamma_{m,n}$, $\Gamma_{m,o}$ and $\Gamma_{n,o}$ are of type 2, possessing the interval $I_{mno} := [m \wedge n \wedge o, m \vee n \vee o]$. From setting (6.4) we conclude $U_m(I_{mno}) < U_n(I_{mno}) < U_o(I_{mno})$. This is, $U_m(I_{mno}) < U_o(I_{mno})$ and according to (6.4) $m \, \hat{L}_{m,o} \, o$. $\square$

By turning components due to the changed orientations of $\hat{L}$ in Lemma 6.34 we gain a bipartition of Γ that provides a left-order on the respective lattice $\mathfrak{V}$. This is the first main result of this chapter.

Theorem 6.35 *[Zsc07a] Let $\mathbb{K} = (J, M, \leq)$ be the standard context of a lattice $\mathfrak{V}$ and Γ its Ferrers-graph. Then*

$$\Gamma \text{ is bipartite} \iff \mathfrak{V} \text{ is planar.}$$

Proof: "$\Rightarrow$": Consider the relation $\hat{L}$ from Lemma 6.34. Let $\hat{\mathfrak{L}} := \bigcup_{k \in K} \hat{\mathfrak{L}}_k$ and $\hat{\mathfrak{R}} := \bigcup_{k \in K} \hat{\mathfrak{R}}_k$ with
$$\hat{\mathfrak{L}}_k := \begin{cases} \mathfrak{L}_k, & \hat{L}_k = \tilde{L}_k \\ \mathfrak{R}_k, & \hat{L}_k = \tilde{L}_k^{-1} \end{cases}.$$

$\hat{\mathfrak{L}}$ and $\hat{\mathfrak{R}}$ are a bipartition of Γ since $\mathfrak{L}_k = V_k(\Gamma) \cap \mathfrak{L}$ (see Remark 6.33) and $\hat{L}$ is its induced relation. The relation $\hat{L}$ is asymmetric and connex (Lemma 6.4) and transitive (Lemma 6.34). Therefore $\mathfrak{V}$ is planar (Lemma 6.5).

"⇐": If $\mathfrak{V}$ is planar then its order dimension is at most two (Theorem 3.20). Therefore, the Ferrers-dimension is at most two (Theorem 3.39) and hence its Ferrers-graph bipartite (Lemma 3.42). □

In this chapter we used to consider standard contexts $\mathbb{K}$ of lattices until now. This seems reasonable since we were interested in a lattice property (namely planarity) that we were able to derive from a certain property of $\mathbb{K}$ (namely the bipartiteness of the associated Ferrers-graph). Having a look on FCA as applied lattice theory, one can ask now, whether we must restrict us to *reduced* contexts as equivalent to $\mathbb{K}$ or whether it suffices to consider any context to check the planarity of the appropriate concept lattice. We will see that the second is true.

As a preparation, we first prove that inserting or deleting rows and columns in a context $\mathbb{K}$ do not affect the bipartiteness of its Ferrers-graph Γ if $\mathfrak{B}(\mathbb{K})$ remains the same up to isomorphy.

Lemma 6.36 *Let $\mathbb{K} = (G, M, I)$ be a context possessing the Ferrers-graph Γ and $\tilde{\mathbb{K}}$ be the reduced context of $\mathbb{K}$ possessing the Ferrers-graph $\tilde{\Gamma}$. Then Γ is bipartite if and only if $\tilde{\Gamma}$ is bipartite.*

Proof: "⇒": $\tilde{K}$ emerges from $\mathbb{K}$ by deleting some rows and columns. For the appropriate Ferrers-graphs, we note that $\tilde{\Gamma}$ results from Γ by deleting certain vertices and their incident edges. Clearly, the property of bipartiteness is conserved.

"⇐": Let n be a non-$\bigwedge$-irreducible element of M. Then $\mu n = \bigwedge M_n$ for a set $M_n \subseteq M$. Now let $\{(g, n), (h, m)\} \in E(\Gamma)$. There exists a $\bigwedge$-irreducible $m_1 \in M_n$ with $g \not I m_1$. Since $m' \subseteq m_1'$, we conclude also hIm_1 and hence $\{(g, m_1), (h, m)\} \in E(\Gamma)$. We conclude $N(g, m) \subseteq N(g, m_1)$, where $N(\gamma)$ denotes the set of neighbors of a vertex $\gamma \in \Gamma$. Therefore, for every (odd) circle in the Ferrers-graph Γ we find another one in $\tilde{\Gamma}$. □

Applying this lemma we may extend the characterization given in the main result Theorem 6.35 to arbitrary contexts and their respective concept lattices. More precisely, a context $\mathbb{K}$ with bipartite Ferrers-graph possesses a planar concept lattice $\mathfrak{B}(\mathbb{K})$ and vice versa all contexts $\tilde{\mathbb{K}}$ having a concept lattice isomorphic to the planar one $\mathfrak{B}(\mathbb{K})$ possess a bipartite Ferrers-graph.

Corollary 6.37 *Let $\mathbb{K} = (G, M, I)$ be a context with Ferrers-graph Γ. Then*

$$\Gamma \text{ is bipartite} \iff \mathfrak{B}(\mathbb{K}) \text{ is planar}.$$

Proof: Follows from Lemma 6.36, Theorem 6.35 and the fact that for an arbitrary context $\mathbb{K}$ and its reduced version $\tilde{\mathbb{K}}$, the equality $\mathfrak{B}(\mathbb{K}) \cong \mathfrak{B}(\tilde{\mathbb{K}})$ holds. □

6.3.2 The Number of Plane Diagrams of a Lattice

After giving a possibility to find one plane diagram of a lattice now we want to find all of them. Of course, this can be done up to similarity only. One obviously obtains infinitely many plane diagrams by jiggling slightly a node.

Let us recall where we stand: Every plane diagram uniquely (up to similarity) determines a left-order that gives a bipartition on the Ferrers-graph Γ. Vice versa there exist bipartions of Γ inducing left-orders, i.e. plane diagrams. We want to find now all these bipartitions.

Doignon et al. [DDF84] tried to tackle the issue of finding the number of plane diagrams of a poset with the same approach but could not find a general solution. Golumbic [Gol80] solves the problem by counting all transitive orientations[3] of the incomparability graph. However, as far as we know, we are the first to publish an algorithm for this task in Section 6.4.

In the following two lemmas we investigate, in which way type 1 and type 2 components can be turned around while keeping the induced relation transitive. For the former, there are no constraints at all, as the first observation claims. See also Figure 6.14 for a visual substantiation.

Lemma 6.38 *Let $\mathfrak{V}$ be a lattice with bare Ferrers-graph Γ. Let $V(\Gamma) = \mathfrak{L} \stackrel{.}{\cup} \mathfrak{R}$ be a bipartition of Γ, s.t. the induced relation $\tilde{L}_1$ can be extended to a left-order L_1. Furthermore let $\Gamma_\circ$ be a component of Γ of type 1. Then the bipartition*

$$V(\Gamma) = ((\mathfrak{L} \setminus \mathfrak{L}_\circ) \cup \mathfrak{R}_\circ) \stackrel{.}{\cup} ((\mathfrak{R} \setminus \mathfrak{R}_\circ) \cup \mathfrak{L}_\circ)$$

is again a bipartition inducing a relation $\tilde{L}_2$ that can be extended to a left-order L_2 distinct from L_1.

Proof: We have to evidence the following claims

1. $((\mathfrak{L} \setminus \mathfrak{L}_\circ) \cup \mathfrak{R}_\circ) \stackrel{.}{\cup} ((\mathfrak{R} \setminus \mathfrak{R}_\circ) \cup \mathfrak{L}_\circ)$ is again a bipartition of the vertex set of Γ: This is intuitively clear since we just "turned around" one of its components, see Remark 6.33.

2. The induced relation $\tilde{L}_2$ can be extended to a left-order: We know (see Lemma 6.4) that $\tilde{L}_2$ is asymmetric and connex since it is induced by a bipartition. Now assume $\tilde{L}_2$ not to be transitive. Hence we have three $\bigwedge$-irreducibles satisfying $m\ \tilde{L}_2\ n\ \tilde{L}_2\ o\ \tilde{L}_2\ m$. Since $\tilde{L}_1$ is transitive by precondition, let w.l.o.g. $m\ \tilde{L}_1\ n$. In particular this means $\Gamma_\circ = \Gamma_{m,n}$. Otherwise we know by Corollary 6.17 that (m, n, o) is a free triple and, due to Lemma 6.30, that $\Gamma_\circ$ is of type 2. This contradicts our precondition.

---
[3] A transitive orientation (V, F) is an orientation of a simple undirected graph (V, E), s.t. $F \cap F^{-1} = \emptyset$, $F \cup F^{-1} = E$ and $F \circ F \subseteq F$ hold [Gol80].

106   Chapter 6: Left-relations on Contexts

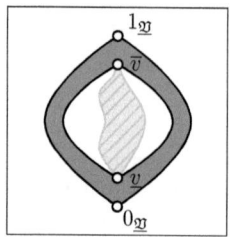 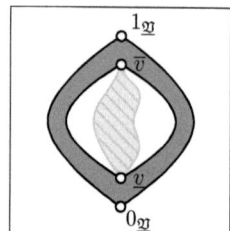

Figure 6.14: In a plane diagram, turning a type 1 component results again in a plane diagram. The interval $[\underline{v}, \overline{v}]$ of the component (depicted diagonally striped) is thereby reflected.

3. $L_1 \neq L_2$: By Definition 6.26, $\Gamma_\circ$ contains at least one edge. Let $\Gamma_\circ = \Gamma_{m,n}$ for some $m, n \in M$. W.l.o.g. let $m \, \tilde{L}_1 \, n$ and $n \, \tilde{L}_2 \, m$. This extends to the respective left-relations, i.e. $m \, L_1 \, n$ and $n \, L_2 \, m$. Both left-orders are asymmetric, so we conclude $L_1 \neq L_2$. □

Type 2 components have to be handled more carefully. As already suggested in the proof of Lemma 6.34, the orientations of those components with the same interval have to comply with a certain order of the accompanying m-components. Therefore, the number of bipartitions equals to the number of linear orders on the set of respective m-components. See Figure 6.15 for an example.

Lemma 6.39 *[Zsc08] Let $\mathbb{K} = (J, M, \leq)$ be the standard context of a lattice $\mathfrak{V}$ with Ferrers-graph Γ. Let $V(\Gamma) = \mathfrak{L} \, \dot{\cup} \, \mathfrak{R}$ be a bipartition of Γ, s.t. the induced relation $\hat{L}$ is transitive. Furthermore let $\Gamma_{[\underline{v},\overline{v}]} := \{\Gamma_1, \ldots, \Gamma_r\}$ be the set of components of type 2 possessing the interval $[\underline{v}, \overline{v}]$ and s the number of m-components in $[\underline{v}, \overline{v}]$, i.e. $U([\underline{v},\overline{v}]) = U_{m_1}[\underline{v},\overline{v}], \ldots, U_{m_s}[\underline{v},\overline{v}]$. Then*

1. $r = \binom{s}{2}$.

2. *Let $K \subseteq \{1, \ldots, r\}$. The relation $\hat{L}$ induced by the bipartition*

$$V(\Gamma) = \left(\left(\mathfrak{L} \setminus \bigcup_{k \in K} \mathfrak{L}_k\right) \cup \bigcup_{k \in K} \mathfrak{R}_k\right) \dot{\cup} \left(\left(\mathfrak{R} \setminus \bigcup_{k \in K} \mathfrak{R}_k\right) \cup \bigcup_{k \in K} \mathfrak{L}_k\right) \quad (6.5)$$

is transitive if and only if it is a linear order on the set $M_s = \{m_1, \ldots, m_s\}$.

3. *There exist exactly $s!$ bipartitions of the form of Equation (6.5) whose induced relations are transitive.*

6.3 Gaining Left-orders out of Ferrers-graphs

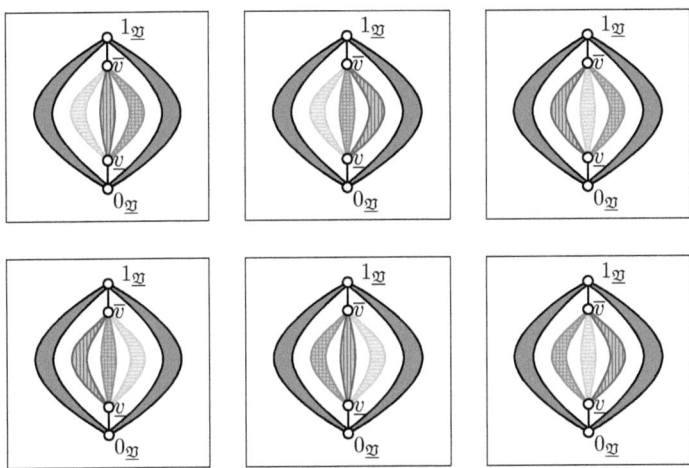

Figure 6.15: In a plane diagram, turning type 2 components results again in a plane diagram if their respective ordering corresponds to a linear ordering of the involved proper components. In this picture we have three proper components U_{m_1} (horizontally striped), U_{m_2} (vertically striped) and U_{m_3} (crosshatched). Therefore there exist three components of type 2, namely Γ_{m_1,m_2}, Γ_{m_1,m_3} and Γ_{m_2,m_3}. Of the eight possible orientations of that components, six supply a plane diagram.

Proof:

1. Due to Lemma 6.30 there exist (at least) $\binom{s}{2}$ different components Γ_{m_i,m_j} (with $i \neq j \in \{1,\ldots,s\}$) possessing the interval $[\underline{v},\overline{v}]$.

   Due to Corollary 6.27 each component Γ_{m_i,m_j} possessing the interval corresponds to a pair of m-components $(U_{m_i}([\underline{v},\overline{v}]), U_{m_j}([\underline{v},\overline{v}]))$. Hence, $r = \binom{s}{2}$.

2. Due to Definition 6.12 and Lemma 6.4 we know that $\hat{L}$ is asymmetric and connex on M_s. If $\hat{L}$ is not a linear order on M_s then it is not transitive on M_s and therefore not transitive on M.

   Now assume to find a triple $n_1, n_2, n_3 \in M$ satisfying $n_1 \hat{L} n_2 \hat{L} n_3 \hat{L} n_1$. By Corollary 6.17 we know that (n_1, n_2, n_3) is a free triple. Since $\tilde{L}$ is transitive and $\hat{L}$ evolved as the induced relation of the bipartition (6.5) where we just turned around some components of $\Gamma_{[\underline{v},\overline{v}]}$, let w.l.o.g.

$\Gamma_{n_1,n_2} = \Gamma_{m_1,m_2}$. We conclude that the components Γ_{n_1,n_2}, Γ_{n_1,n_3} and Γ_{n_2,n_3} are of type 2 (Lemma 6.30) possessing the interval $[\underline{v},\overline{v}]$. Hence, the m-components $U_{n_i}([\underline{v},\overline{v}])$ are pairwise disjoint (Corollary 6.27). W.l.o.g. let $n_i \in U_{m_i}([\underline{v},\overline{v}])$ for $i \in \{1,2,3\}$. With Lemma 6.32 we conclude $m_1 \hat{L} m_2 \hat{L} m_3 \hat{L} m_1$, i.e. $\hat{L}$ is not a strict order on M_s.

3. This is obvious: there exist $s!$ linear orders on an s-elemental set and therefore also $s!$ orientations of the involved components of Γ inducing a relation $\hat{L}$ that is transitive.

□

Kelly and Rival in [KR75] first observed that flipping or interchanging m-components $U_m([\underline{v},\overline{v}])$ (or precisely the set $\{\bigwedge N \mid N \subseteq U_m([\underline{v},\overline{v}])\}$, they called this a proper $\langle \underline{v},\overline{v}\rangle$-component) keeps a lattice diagram planar. Our achievement is to be able to count the number of those transformations correctly[4] by assigning components of the corresponding Ferrers-graph to m-components (or pair of them). This is the concern of the next result which is the second goal of this chapter.

Theorem 6.40 *[Zsc08] Let $\mathbb{K} = (J, M, \leq)$ be the standard context of a lattice $\mathfrak{V}$ with Ferrers-graph Γ.*

1. *If Γ is not bipartite then $\mathfrak{V}$ possesses no plane diagram.*

2. *Let Γ is bipartite with κ components of type 1 and $n = n_1 + \ldots + n_t$ components of type 2, s.t. $n_i > 0$ is the number of components of the set $\Gamma_{[\underline{v},\overline{v}]}$ for some interval $[\underline{v},\overline{v}]$. Then $\mathfrak{V}$ possesses $2^\kappa \cdot \prod_{i=1}^{t} \mu_i!$ non-similar plane diagrams. Thereby $n_i = \binom{\mu_i}{2}$.*

Proof:

1. This is the assertion of Theorem 6.35.

2. Due to Theorem 6.35 there exists at least one plane diagram. The claim follows then from Lemma 6.38 and Lemma 6.39. Recall that non-similar plane diagrams can be represented by different left-orders and vice versa (see Corollary 5.7).

□

A statement concerning the number of transitive orientations of a comparability graph which is equal to the number of plane diagrams of a poset (see Theorem 3.20) is given in [Gol80]. Although relying on different constructions, a product of factorials as resulting number is the outcome, too.

---
[4]The problem is, flipping one m-component may result in the same diagram as interchanging two others.

6.3 Gaining Left-orders out of Ferrers-graphs

See Figure 6.16 for an example. We consider the lattice we already introduced in Figure 6.10. There we noticed that the Ferrers-graph consists of two components, one being of type 1 and the other one of type two. Since $1 = \binom{2}{2}$, we conclude with Theorem 6.40 that the lattice possesses $2^1 \cdot 2! = 4$ plane diagrams. These are depicted in the Figure. One may observe that flipping the type 1 component $\Gamma_{n,o}$ results in turning around its m-component (more precisely, the set $\{\bigwedge N \mid N \subseteq U_n(I(\Gamma_{n,o}))\}$). However, flipping the type 2 component $\Gamma_{m,n}$ results in interchanging its m-components, i.e. the sets $\{\bigwedge N \mid N \subseteq U_m(I(\Gamma_{n,o}))\}$ and $\{\bigwedge N \mid N \subseteq U_n(I(\Gamma_{n,o}))\}$.

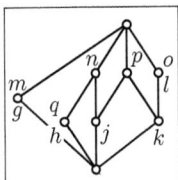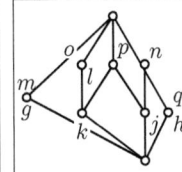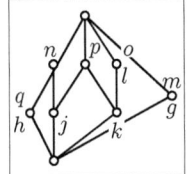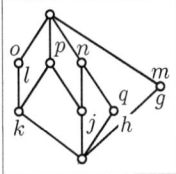

Figure 6.16: All plane diagrams (up to similarity) of a lattice.

After giving the number of non-similar plane diagrams, we also want determine the number of realizers of a planar lattice. As Lemma 3.18 may suggest already, the number of left-orders equals twice the number of realizers since every conjugate order together with its inverse uniquely determines a realizer and vice versa every realizer ascertains two conjugate orders L and L^{-1}.

Corollary 6.41 *Let $\mathfrak{V}$ be a lattice with the bipartite bare Ferrers-graph Γ as previously described in Theorem 6.40. Then $\mathfrak{V}$ possesses $2^{\kappa-1} \cdot \prod_{i=1}^{t} \mu_i!$ realizers.*

Proof: The number of plane diagrams of $\mathfrak{V}$ equals to the number of conjugate orders of $\mathfrak{V}$, i.e. to half the number of realizers of $\mathfrak{V}$ (see Corollary 5.7, Theorem 4.9 and Lemma 3.18). The claim follows with Theorem 6.40. □

In [DDF84] a characterization of posets of dimension 2 possessing a unique realizer is given. Applying our consideration we can derive this result, restricted to lattices, too:

Corollary 6.42 *[DDF84] Let $\mathfrak{V}$ be a lattice with the bipartite bare Ferrers-graph Γ. Then $\mathfrak{V}$ possesses a unique realizer of size two if and only if Γ is connected.*

Proof: Follows immediately from Corollary 6.41. □

Finally we can specify the subset of natural numbers that can be described as the number of non-similar plane diagrams of a lattice:

Chapter 6: Left-relations on Contexts

Corollary 6.43 *Let α be a natural number. There exists a lattice possessing α non-similar plane diagrams if and only if $\alpha = 0$ or $\alpha = \prod_{i=1}^{t} \alpha_i!$ for natural numbers $\alpha_1, \ldots, \alpha_t$.*

Proof: The case $\alpha = 0$ refers to non-planar lattices. The number of non-similar plane diagrams of a planar lattice is a product of factorials; this is implied by Theorem 6.40 if one recalls $2 = 2!$.

Otherwise for any number $\alpha = \prod_{i=1}^{t} \alpha_i!$, the lattice consisting of a parallel composition of lattices M_{α_i} has exactly α plane diagrams: The bare Ferrers-graph of the lattice M_n[5] possesses $\binom{n}{2}$ components of size 2, which all share the interval $[0_{M_n}, 1_{M_n}]$ and has therefore $n!$ plane diagrams. Additionally, the bare Ferrers-graph Γ of the parallel composition of two lattices $\mathfrak{V}_1$ and $\mathfrak{V}_2$ is exactly the (disjoint) union of the bare Ferrers-graphs of its parts. This is due to the fact that all pairs of the form (j, m) with $j \in J(\mathfrak{V}_1)$ and $m \in M(\mathfrak{V}_2)$ or $j \in J(\mathfrak{V}_2)$ and $m \in M(\mathfrak{V}_1)$ fulfill either $j \leq m$ or $j \geq m$. Both cases imply $(j, m) \notin V(\Gamma)$. $\square$

See Figure 6.17 for an example of the previous result. On the left we see the ordinal sum $M_3 + M_3$ of two lattices M_3. Obviously, there exist $(3!)^2 = 36$ non-similar plane diagrams of this lattice: By permuting the $\wedge$-irreducibles a, b and c one gains 6 possibilities and while permuting d, e and f the other 6.

One may argue that all the different diagrams are isomorphic, i.e. for each pair of those diagrams given by the mappings pos_1 and pos_2 we can find an isomorphism φ of $M_3 + M_3$, s.t. both diagrams are equal, i.e. $\text{pos}_1 = \text{pos}_2 \circ \varphi$. However, by adding some extra nodes making the branches distinguishable, non-similar diagrams are not isomorphic in this sense anymore.

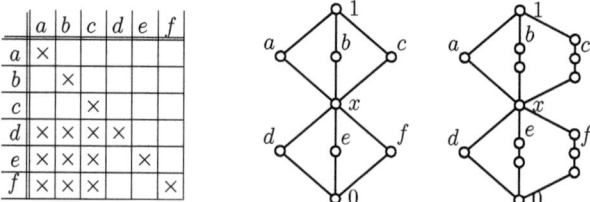

Figure 6.17: Two lattices possessing 36 non-similar plane diagrams.

---

[5]The lattice M_n consists of an n-elemental antichain completed by top and bottom element.

6.4 Determining All Non-similar Plane Diagrams of a Lattice

With the help of the previous statements we can design an algorithm that specifies all left-orders, i.e. all non-similar plane diagrams, of a lattice $\mathfrak{V}$. it consists of the following steps:

1. Calculate the bare Ferrers-graph Γ of $\mathfrak{V}$.

2. Decide, whether Γ is bipartite. If yes, assign a bipartition $\mathfrak{L} \dot{\cup} \mathfrak{R}$ to the vertices of Γ.

3. Determine all components of Γ.

4. Calculate the interval $I(\Gamma_o)$ of each component Γ_o and determine, whether Γ_o is of type 1 or type 2.

5. Find a partition of the set of components of Γ into a set of components of type 1 and sets of components of type 2 sharing the same interval:

$$\Gamma = \{\Gamma_o \mid \Gamma_o \text{ is of type } 1\} \dot{\cup} \bigcup_{\underline{v} < \overline{v} \in \mathfrak{V}} \Gamma_{[\underline{v},\overline{v}]}$$

6. Calculate the m-components of each interval $[\underline{v}, \overline{v}]$ defining a non-empty set $\Gamma_{[\underline{v},\overline{v}]}$.

7. Find all bipartitions inducing a relation $\tilde{L}$ that can be extended to a left-order.

The algorithm finds exactly all bipartitions that define a left-order. This is assured by Theorem 6.40. To gain a plane diagram from that, one can use for instance the method described in Section 5.3.

The last step of the algorithm can not be processed in polynomial time (both in terms of $\bigwedge$-irreducibles and of lattice elements). This is due to the fact that a planar lattice may possess up to $|M(\mathfrak{V})|!$ non-similar plane diagrams. Actually it is not possible to write them down or draw them in polynomial time. However, the other steps are polynomial however, as we want to clarify in the following. In particular, just determining the *number* of plane diagrams of a lattice can be found in polynomial time.

1. The first step needs a complexity of $\mathcal{O}((|J| \cdot |M|)^2)$ by a naive calculation.

2. Finding a bipartition of a graph (V, E) can be done in $\mathcal{O}(|E|) = \mathcal{O}((|J| \cdot |M|)^2)$ by a breadth-first-search algorithm [Jun94].

3. Finding all components of the bare graph can be done in $\mathcal{O}(|E|) = \mathcal{O}((|J|\cdot|M|)^2)$ by a variation of a breadth-first-search algorithm [Jun94].

4. In a planar lattice every element v can be represented as the infimum of at most two $\bigwedge$-irreducibles and as the supremum of at most two $\bigvee$-irreducibles (see Lemma 3.26). Therefore the calculation of the intervals can be done in $\mathcal{O}(|M|^2)$.

   The distinction between type 1 and type 2 components can be done easily in this process, as Lemma 6.28 states: In a component $\Gamma_\circ$ of type 2, every pair of $\bigwedge$-irreducibles $m, n \in M(\Gamma_\circ)$ has the same infimum $m \wedge n = \underline{v}$ (where $I(\Gamma_\circ) = [\underline{v}, \overline{v}]$). In contrast, if $\Gamma_\circ$ is of type 1 then there are two $\bigwedge$-irreducibles $m, n \in M(\Gamma_\circ)$ satisfying $m \wedge n > \underline{v}$.

5. This step is obviously done in $\mathcal{O}(n)$ where $n \leq |M|^2$ is the number of components of type 2.

6. Exactly those pairs of $\bigwedge$-irreducibles m, n that are connected by an edge of the form $\{(j, m), (h, n)\}$ are in different m-components (see Lemma 6.32). That is, every component $\Gamma_{m,n}$ of type 2 defines two m-components by the given unique bipartition $V(\Gamma_{m,n}) = \mathfrak{L} \dot\cup \mathfrak{R}$ by

$$U_m(I(\Gamma_{m,n})) = \{\tilde{m} \in M(\Gamma_{m,n}) \mid \nexists j, h \in J : \{(j, \tilde{m}), (h, m)\} \in E(\Gamma_{m,n})\}$$
$$U_n(I(\Gamma_{m,n})) = \{\tilde{m} \in M(\Gamma_{m,n}) \mid \exists j, h \in J : \{(j, \tilde{m}), (h, m)\} \in E(\Gamma_{m,n})\}$$

Hence this step takes a complexity of $\mathcal{O}(|M(\Gamma_\circ)|)$ for each component $\Gamma_\circ$ and $\mathcal{O}(|M|^3)$ altogether.

By this consideration we observe that the first six steps of the algorithm need a time complexity of

$$\max\{\mathcal{O}((|J|\cdot|M|)^2), \mathcal{O}(|M|^3), \mathcal{O}(|M|^2)\} \leq \max\{\mathcal{O}((|J|^4), \mathcal{O}(|M|^4)\} = \mathcal{O}(|\mathfrak{V}|^2).$$

See Proposition 3.27 for the equality between the second and third term. The number of plane diagrams can be calculated already after accomplishing step 5 by the formula

$$2^\kappa \cdot \prod_{\Gamma_{[\underline{v},\overline{v}]} \neq \emptyset} \mu_{[\underline{v},\overline{v}]},$$

where $\kappa = |\{\Gamma_\circ \mid \Gamma_\circ \text{ is of type 1}\}|$ is the number of type 1 components and $\mu_{[\underline{v},\overline{v}]}$ is the number of m-components of an interval $[\underline{v}, \overline{v}]$ if that interval possesses several m-components at all, i.e. $|\Gamma_{[\underline{v},\overline{v}]}| = \binom{\mu_{[\underline{v},\overline{v}]}}{2}$.

Let us finally have another look at the last step. "Writing down" all left-orders here means to provide all possible linear orders on the sets of m-components sharing the same upper neighbor and to arrange the components

6.4 Determining All Non-similar Plane Diagrams of a Lattice

of the Ferrers-graph Γ respectively. Finding all linear orders (or, equivalently, all permutations) of an n-elemental set can not be done in polynomial time, since its number is not a polynom of n, but $n!$ instead.

However, determining them can be done in *polynomial delay*. That means, it takes only polynomially many steps to calculate each configuration. There is a well known algorithm called NEXT PERMUTATION [Knu05] finding all permutations of an n-elemental set with linear delay. Hence, specifying all the $\alpha \leq |M|!$ plane non-similar diagrams of a lattice can be done in $\mathcal{O}(|M| \cdot \alpha)$.

Chapter 7

Conclusion and Further Work

7.1 Conclusion

In this section we want to highlight, what we achieved in this work. We recapitulate the most important conclusions and compare them with existing results.

Left-relations on Lattices

We found an efficient way to describe conjugate orders by introducing left-relations on lattices. They can be uniquely represented by sorting relations (on lattices). This means in particular that not the exact positions of the diagram nodes of the $\bigwedge$-irreducible elements, but only their relative positions to each other determine a lattice diagram to be plane or not.

The importance of the $\bigwedge$-irreducibles (and dually $\bigvee$-irreducibles) for the planarity of the underlying lattice becomes evident once again in the planarity conditions. Particularly, the FPC is a helpful result that simplifies many proofs if we know a left-relation not completely.

Finally, the TPC gives a quite intuitive and easily presentable imagination on how the standard context of a planar lattice looks like. Moreover, a respective enumeration ε provides a sorting relation that can be used to draw a plane diagram.

Left-relations on Diagrams

By investigating this issue we contributed two advances. Firstly, we showed in Theorem 5.6 that every conjugate order uniquely determines a plane diagram up to similarity and vice versa. This fact was, as we think, already known, but never explicitly formulated before. Secondly, by modifying the left-right

7.1 Conclusion

numbering method that emerges from the proof of Theorem 5.6 we were able to create an algorithm that draws plane attribute additive diagrams of planar lattices.

Left-relations on Contexts

In our opinion, the main contribution of this dissertation comes from the final part. By introducing left-relations on contexts in an adequate manner we found a fruitful coherence to Ferrers-graphs. This enabled us to investigate the interplay between properties of these graphs and the shape of the appropriate lattice. Although these correlations itself might be an interesting research issue, we want to discuss the two main results of that chapter only.

In Theorem 6.35 we gave a possibility to characterize planar lattices, i.e. lattices of order dimension at most two. Due to the proof, constructing the Ferrers-graph out of the standard context and checking it for inducing a left-relation has a time complexity of $\mathcal{O}(|V|^2)$.

It is well known that deciding the order dimension is $\mathcal{NP}$-hard for posets [Yan82] if it exceeds two. However, there already exist many polynomial approaches in case of the dimension being at most two.

For posets there are two algorithms based on procedures to recognize comparability graphs. The one by Golumbic [Gol80] has a complexity $\mathcal{O}(\delta \cdot |E|)$ while the other by Spinrad and McConnell [MS97] is quicker with complexity $\mathcal{O}(|V| + |E|)$. Thereby, V and E denote the number of vertices and edges of the comparability graph of a poset and δ its maximal degree[1]. Both methods are not quicker than the one we developed in this work, but may be applied on a broader range since they are designed more generally for posets. As well as we did, Doignon et al. [DDF84] tackles the problem with the help of Ferrers-graphs. However, his result is not constructive.

If we consider lattices in particular, the result of Platt (see Theorem 3.23) supplies together with a quick graph planarity algorithm (e.g. [HT74], [LEC67]) a possibility to recognize and lay out planar lattices with a complexity of $\mathcal{O}(|V|)$. Hence, it is the quickest available method so far.

In the other result in Theorem 6.40 we give an algorithm to calculate the number of all plane diagrams in $\mathcal{O}(|V|^2)$. Actually writing down the corresponding left-orders can be done with polynomial delay, namely $\mathcal{O}(|M|)$. In the worst case - if the observed lattice is of the form M_n - calculating all $n!$ left-orders has a complexity of $\mathcal{O}(|M| \cdot n!)$ therefore. An algorithm as described in Section 6.4 was, as far as we know, not known yet.

---

[1] In graph theory, the *degree* of a vertex v is the number of its neighbors, i.e. the number of vertices adjacent to v.

7.2 Further Work

7.2.1 Algebraic Consequences

Of course, this dissertation aims at an applicational direction. The results may be used primarily for designing better algorithms for drawing diagrams of lattices. However, some of the newly introduced concepts may deserve to be investigated further.

One the one hand one can observe left-orders more deeply. Questions could include the following: Is it possible to characterize those (planar) lattices whose conjugate order is a lattice itself? Which lattices (or posets) have conjugate orders isomorphic to the original order? Which lattices/posets can be represented as conjugate orders?

On the other hand it seems to be fruitful also to analyze Ferrers-graphs further. We only observed the bipartite ones. In the general case, the partition into *types* according to Definition 6.26 remains the same. However, the induced relations on type 1 components are no orders anymore. Contrary, type 2 components always seem to possess an order as induced relation since thay are complete directed bipartite graphs (see Lemma 6.32). We think that the rich coherences between lattices and Ferrers-graphs, between the different concepts of components and of connections give a broad space for further discoveries.

7.2.2 Minimizing the Crossing Number

Although it was the initial concern of this work to find strategies to minimize the crossing number, the fruitfulness of the prerequisite theory of planarity did not allow us to investigate it reasonably yet. However, we discovered some preliminary results that we present here. Thus, we try to indicate the chances that the theory of Ferrers-graphs introduced in Chapter 6 offers for advancing in this topic.

By the well-known result Garey and Johnson [GJ83] we know that minimizing the crossing number cr of a graph is $\mathcal{NP}$-complete. The same holds for posets and lattices [Fre04]. However, it might be possible that the problem becomes tractable if we choose cr small enough. The crossing number problem for graphs, i.e. deciding whether a simple undirected graph has a diagram with at most cr edge crossings, can be solved in $\mathcal{O}(|V|^2)$ time by an algorithm of Grohe [Gro01]. However, the constant factor grows doubly exponentially in cr. Hence, it is efficient if cr is small, only. A problem that can be solved - for a fixed parameter k - in polynomial time is called *fixed-parameter-tractable* [FG06].

7.2 Further Work

There are several approaches to describe the "non-planarity" of lattices (or, more general, of graphs). One is by means of the above-mentioned crossing number, another by the size of a minimal subset $W \subseteq V$ whose deletion makes the remaining induced subgraph planar. We present a new approach (for lattices only) that is inspired by the usefulness of Ferrers-graphs.

Definition 7.1 *Let $\mathfrak{V}$ be a lattice with standard context (J, M, I). We call $\mathfrak{V}$ ∇-naplar if there exists a planar lattice $\mathfrak{W}$ generated by the context $(J, M, \tilde{I})$ meeting[2] $|I \triangle \tilde{I}| = \nabla$.* ◊

In particular we will be interested in the simplest case of 1-naplar lattices. They are characterized by a standard context that becomes planar (in the sense of Section 4.4) by deleting or adding a cross. A first result on nearly planar lattice is the following observation which is proved in [Zsc07c].

Lemma 7.2 *Let $\mathbb{K} = (G, M, I)$ be a context and $\Gamma(I)$ its Ferrers-graph. Furthermore, let the chromatic number $\chi(\Gamma(I)) = 3$ and $\bar{I} = I_1 \stackrel{.}{\cup} I_2 \stackrel{.}{\cup} I_3$ a valid partition into color classes with $|I_1| = 1$. Then, $\Gamma(I \cup I_1)$ is bipartite.*

That is, if we have a context whose empty cells can be tricolored with a valid vertex coloring (e.g. with black, light and shaded), s.t. one color (e.g. black) is used exactly once then replacing the appropriate node by a cross leaves the Ferrers-graph 2-colorable. This sounds obvious. In fact, adding a cross in $\mathbb{K}$ also adds new edges in the graph $\Gamma(I)$, i.e. the modified $\Gamma(I \cup I_1)$ is not just the subgraph induced by $I_2 \cup I_3$. See picture for an example, replacing the black dot by a cross adds the edge $\{(h, a), (g, b)\}$.

	a	b	c	d
g	×	○	○	○
h	○	●	×	×
j	○	×	○	×
k	○	×	×	○

We may conclude that every lattice possessing a standard context satisfying the preconditions of Lemma 7.2 is 1-naplar. This allows us to ask for the crossing number of certain lattices possessing a Ferrers-graph that is "nearly" bipartite. Although we did not prove the following conjecture yet, for small lattices it seems to be true.

Conjecture: Let $\mathfrak{V}$ be a lattice. Let the Ferrers-graph of $\mathfrak{V}$ possess exactly one chordless cycle of odd length $2n + 1$. Then $cr(\mathfrak{V}) = \begin{cases} 2, & n = 1 \\ 1, & n > 1 \end{cases}$.

---
[2] By $\triangle$ we denote the operator of the symmetric difference.

Unfortunately, 1-naplar lattices may have an arbitrary crossing number in general. Consider the lattice $M_r \times D_2$ as an example. Its Ferrers-graph consists of a set of triangles with the vertices (g, m), (h_i, n_j) and (h_j, n_i) (for all $i \neq j \in \{1, \ldots, r\}$). Removing the black dot leaves the Ferrers-graph bipartite. However, the lattice has (with growing r) an increasing number of edge crossings.

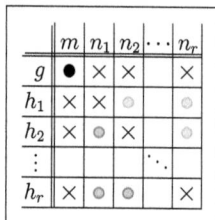

A long-term objective of all this considerations could be to characterize lattices having a small crossing number cr by their respective Ferrers-graphs, e.g. by the number and length of their chordless odd cycles. This could be possible even in polynomial time. In order to figure out that a graph possesses at most n odd cycles one has to delete all n-elemental vertex subsets respectively and to check the remaining induced subgraph to be bipartite. This requires a complexity of $\mathcal{O}(|V|^n \cdot |V^2|)$. That means that the problem becomes fixed-parameter tractable. Furthermore, it ought to be possible to generate a left-relation that describes a diagram possessing cr edge crossings.

7.2.3 Drawing Nice Diagrams

Recent lattice layout algorithms do not include strategies for recognizing or drawing planar lattices. Only methods based on layer assignment use heuristical approaches for minimizing the crossing number. Our framework does not only allow to easily implement a procedure for detecting and drawing planar lattices (a feature that has not been used yet since not very many lattices in practice are planar indeed), but the possibility to find all plane diagrams up to similarity offers a new approach for drawing non-planar structures as well. Given a lattice $\mathfrak{V}$ with standard context $\mathbb{K}$, this can be done by performing the following steps:

1. Find, by deleting some $\bigwedge$-irreducibles , all maximal subcontexts that possess a planar lattice.

2. Find, for each subcontext found in step 1., all plane diagrams of the respective lattice.

3. Add the deleted $\bigwedge$-irreducibles into each context found in step 1., i.e. add the respective prime ideals into the diagrams of step 2., s.t. the *conflict distance*[3] is being maximized.

---

[3]That is, the least distance between a diagram node and a non-incident diagram edge

7.2 Further Work

4. Apply a *force directed placement*[4] algorithm on each diagram found in step 3. to further enhance the conflict distance and to fulfill other esthetic criteria.

5. Apply a quality function that chooses the best diagram (or diagrams) due to a certain metrics.

The first step is to be understood that a minimal number of $\bigwedge$-irreducibles should be removed.

Step 3 can be undertaken by using an algorithm proposed by Schmidt [Sch02] that assigns a vector to a $\bigwedge$-irreducible m maximizing the conflict distance between the original diagram and the newly included prime ideal of m.

Step 4 demands some more experience with the importance of esthetic criteria and some intuition about how to create forces to model them, of course. Many algorithms exist already (see, e.g. [Ead84, FR91] or [DETT99] for an overview of the topic), although most of them are designed for undirected graphs only. However, we have our own framework [Zsc07b] that could be extended in an appropriate manner.

The last step seems to be the most complicated. One has to balance the importance of the included esthetic criteria quantitatively in order to weight them. A naive formula could be of the form

$$q(cr, dst) := \alpha \cdot cr + \beta \cdot dst^{-1}.$$

Thereby the quality function q has to be minimized. The crossing number cr is supposed to be small as well as the inverse of the conflict distance dst. The parameters α and β have to be chosen in a way that indeed the best diagram is selected.

---

[4] An algorithm class developed in the graph drawing community. A diagram is considered as a physical system with forces that act on vertices and edges due to the desired diagrams properties. A balanced state of the system often supplies a good layout.

Bibliography

[BCB+92] P. Bertolazzi, R. Cohen, G. Di Battista, R. Tamassia, and I. Tollis, *How to draw a seies-parallel digraph*, Algorithm Theory - SWAT '92, LNCS 621, 1992, pp. 272–283.

[BFR71] K. A. Baker, P.C. Fishburn, and F. S. Roberts, *Partial orders of dimension 2*, Networks **2** (1971), 11–28.

[BHS02] P. Becker, J. Hereth, and G. Stumme, *ToscanaJ - an open source tool for qualitative data analysis*, Proc. of FCAKDD 2002, 2002, pp. 1–2.

[Bir67] G. Birkhoff, *Lattice theory*, third ed., Amer. Math. Soc., 1967.

[BL76] K. S. Booth and G.S. Lueker, *Testing for the consecutive one property, interval graphs, and graph planarity using PQ-tree-algorithms*, Journal of Computer and System Sciences **13** (1976), 335–379.

[BLS99] A. Brandstädt, V. Le, and J. Spinrad, *Graph classes - a survey*, SIAM, 1999.

[CDE06] R.J. Cole, J. Ducrou, and P. Eklund, *Automated layout of small lattices using layer diagrams*, Proc. of ICFCA06, LNAI 3874, 2006, pp. 291–305.

[Cog82] O. Cogis, *On the ferrers-dimension of a digraph*, Discrete Math. **38** (1982), 47–52.

[Col00] R.J. Cole, *Automated layout of concept lattice using layer diagrams and additive diagrams*, Proc. of 24. Australasian Comp. Sc. Conf., Australian Comp. Sc. Communications., vol. 23, 2000, pp. 47–53.

[CS00] R.J. Cole and G. Stumme, *CEM: A conceptual email manager*, Proc. of ICCS00, LNAI 1867, 2000, pp. 438–452.

[CT94] I. Cruz and R. Tamassia, *Tutorial on graph drawing*, 1994, http://www.cs.brown.edu/people/rt/gd-tutorial.html.

[DDF84]   J.P. Doignon, A. Ducamp, and J.C. Falmagne, *On realizable biorders and the biorder dimension of a relation*, J. of. Math. Psychology **28** (1984), 73–109.

[DE05]   J. Ducrou and P. Eklund, *Combining spatial and lattice-based information landscapes*, Proc. of ICFCA05, LNAI 3403, 2005, pp. 64–78.

[DE07]   _____, *Searchsleuth: The conceptual neighbourhood of a web query*, Proc. of CLA07, 2007, pp. 253–263.

[DETT99] G. DiBattista, P. Eades, R. Tamassia, and I. G. Tollis, *Graph drawing*, Prentice Hall, 1999.

[Die96]   R. Diestel, *Graphentheorie*, Springer, 1996.

[DM41]   B. Dushnik and E.W. Miller, *Partially ordered sets*, Amer. J. Math. **63** (1941), 600–610.

[DP02]   B. Davey and H. Priestley, *Introduction to lattices and order*, second ed., Cambridge Univ. Press, 2002.

[DS05]   A. Das and M. Sen, *Bigraphs/digraphs of Ferrers dimension 2 and asteroidal triples of edges*, Discrete Mathematics **295** (2005), 191–195.

[DSRW89] S. Das, M. Sen, A.B. Roy, and D.B. West, *Interval digraphs: An analogue of interval graphs*, J. Graph Theory **13**, No.2 (1989), 189–202.

[DVE06]   J. Ducrou, B. Vormbrock, and P. Eklund, *FCA-based browsing and searching of a collection of images*, Proc. of ICCS06, LNAI 4068, 2006, pp. 203–214.

[DWE05]   J. Ducrou, B. Wormuth, and P. Eklund, *D-sift: A dynamic simple intuitive FCA tool*, Proc. of ICCS05, LNAI 3596, 2005, pp. 295–306.

[Ead84]   P. Eades, *A heuristic for graph drawing*, Congressus Numerantium **42** (1984), 149–160.

[EDB04]   P. Eklund, J. Ducrou, and P. Brawn, *Concept lattices for information visualization: Can novices read line diagrams?*, Proc. of ICFCA04, LNAI 2961, 2004, pp. 57–72.

[EEH+97] Peter Eades, Hossam Elgindy, Michael Houle, Bill Lenhart, Mirka Miller, and Sue Whitesides, *Dominance drawings of bipartite graphs*, 1997, http://citeseer.ist.psu.edu/614443.html.

[ES90]   P. Eades and K. Sugiyama, *How to draw a directed graph*, J. of Inform. Proc. **13** (1990), 424–437.

[Far48] I. Fary, *On straight line representation of planar graphs*, Acta Sci. Math. (Szeged) **11** (1948), 229–233.

[Fer07] S. Ferre, *Organizing and browsing a personal photo collection with a logical information system*, Proc. of CLA07, 2007, pp. 112–123.

[FG65]   D.R. Fulkerson and O.A. Gross, *Incidence matrices and interval orders*, Pacific Journal of Mathematics **15, no. 3** (1965), 835–855.

[FG06]   J. Flum and M. Grohe, *Parameterized complexity theory*, Springer, 2006.

[FR91]   T. Fruchterman and E. Reingold, *Graph drawing by force directed placement*, Software - Practice and Experience **21, no.11** (1991), 1129–1164.

[Fre]   R. Freese, *Online java lattice building application*, http://maarten.janssenweb.net/jalaba/JaLaBA.pl.

[Fre04] _____, *Automated lattice drawing*, Proc. of ICFCA04, LNAI 2961, 2004, pp. 112–127.

[Gan04] B. Ganter, *Conflict avoidance in order diagrams*, J. of Univ. Comp. Sc. **10, no. 8** (2004), 955–966.

[GH62]   A. Ghouila-Houri, *Charactérisation des graphes non orientés dont on peut orienter les arêtes de manière à obtenir le graphe d'une relation d'ordre.*, C.R. Acad. Sci. Paris **254** (1962), 1370–1371.

[GJ83]   M. R. Garey and D. S. Johnson, *Crossing number is NP-complete*, SIAM J. Alg. Discr. Math. **4** (1983), 312–316.

[Gol80] M. Golumbic, *Algorithmic graph theory and perfect graphs*, Academic Press, 1980.

[Grä98] G. Grätzer, *General lattice theory*, second ed., Birkhäuser, 1998.

[Gro01] Martin Grohe, *Computing crossing numbers in quadratic time*, STOC '01, 2001, pp. 231–236.

[Gut44] L. Guttman, *A basis for scaling qualitative data*, Amer. Sociol. Rev. **9** (1944), 139–150.

BIBLIOGRAPHY

[GW99]   B. Ganter and R. Wille, *Formal concept analysis - mathematical foundations*, Springer, 1999.

[GY99]   J. Gross and J. Yellen, *Graph theory and its applications*, CRC Press, 1999.

[HT74]   J. Hopcroft and R. Tarjan, *Efficient planarity testing*, JACM **21** (1974), 449–568.

[JHSS07] R. Jäschke, A. Hotho, C. Schmitz, and G. Stumme, *Analysis of the publication sharing behaviour in BibSonomy*, Proc. of ICCS07, LNAI 4604, 2007, pp. 283–295.

[Jun94] D. Jungnickel, *Graphen, Netzwerke und Algorithmen*, third ed., BI-Wiss.-Verl., 1994.

[Kel73] D. Kelly, *Planar partially ordered sets*, Tech. report, 1973.

[Kel81] _____, *On the dimension of partially ordered sets*, Discrete Math. **35** (1981), 135–156.

[Kel87] _____, *Fundamentals of planar ordered sets*, Discrete Math. **63** (1987), 197–216.

[Knu05] D. Knuth, *The art of computer programming*, vol. 4, fascicle 2, Addison-Wesley, 2005.

[Kös06] B. Köster, *FooCA: Enhancing google information research by means of formal concept analysis*, Contributions of ICFCA06, 2006, pp. 1–17.

[KR74]   D. Kelly and I. Rival, *Crowns, fences and dismantlable lattices*, Can. J. Math. **26** (1974), 1257–1271.

[KR75]   _____, *Planar lattices*, Can. J. Math. **27** (1975), no. 3, 636–665.

[KT82]   D. Kelly and W. Trotter, *Dimension theory for ordered sets*, Proc. of Symp. on Ordered Sets, Reidel Publishing, 1982, pp. 171–212.

[LEC67] A. Lempel, S. Even, and I. Cederbaum, *An algorithm for planarity testing of graphs*, Theory of Graphs, International Symp., 1967.

[MS97]   R. McConnell and J. Spinrad, *Linear-time transitive orientation*, 8th ACM-SIAM Symp. on Disc. Alg., 1997, pp. 19–25.

[Ore62] O. Ore, *Theory of graphs*, Colloquium Publications, vol. XXXVIII, Amer. Math. Soc, 1962.

[PAC00] H. Purchase, J. Adler, and D. Carrington, *User preference of graph layout aesthetics: A UML study*, Symposium on Graph Drawing GD'00, LNCS 1984, 2000, pp. 5–18.

[Pla76] C.R. Platt, *Planar lattices and planar graphs*, J. Combinatorial Theory Ser. B (1976), 30–39.

[pri] *private communication with B. Ganter.*

[Pur97] H. Purchase, *Which aesthetic has the greatest effect on human understanding?*, Symposium on Graph Drawing GD'97, LNCS 1353, 1997, pp. 248–261.

[Qua73] R.W. Quackenbush, *Planar lattices*, Proceedings of the University of Houston - Lattice Theory Conference 1973 (1973), 512–518.

[Rep07] H. Reppe, *An FCA perspective on n-distributivity*, Proc. of ICCS07, LNAI 4604, 2007, pp. 255–268.

[Reu89] K. Reuter, *Removing critical pairs*, Tech. Report 1241, TU Darmstadt, 1989.

[Rig51] J. Riguet, *Les relations de ferrers*, Comptes renuds des Séances de l'Académie des Sciences **232** (1951), 1729–1730.

[Riv74] I. Rival, *Lattices with doubly irreducible elements*, Can. Math. Bull. **17** (1974), 91–95.

[Sch02] B. Schmidt, *Ein Optimierungsalgorithmus für additive Liniendiagramme*, Master's thesis, TU Dresden, 2002.

[Sko92] M. Skorsky, *Endliche Verbände - Diagramme und Eigenschaften*, Ph.D. thesis, TH Darmstadt, 1992.

[STT81] K. Sugiyama, S. Tagawa, and M. Toda, *Methods for visual understanding of hierarchical system structures*, IEEE Trans. Syst. Man Cybern. **SMC-11, No.2** (1981), 109–125.

[Tro92] W. Trotter, *Combinatorics and partially ordered sets - dimension theory*, John Hopkins Univ. Press, 1992.

[Yan82] M. Yannakakis, *On the complexity of the partial order dimension problem*, SIAM - Alg. Discr. Math. **3** (1982), 351–358.

[Yev] S. Yevtushenko, *Conexp*, http://conexp.sourceforge.net.

BIBLIOGRAPHY

[Zsc05] C. Zschalig, *Planarity of lattices - an approach based on attribute additivity*, Proc. of ICFCA05, LNAI 3403, 2005, pp. 391–402.

[Zsc06a] _____, *Characterizing planar lattices using left-relations*, Proc. of ICFCA06, LNAI 3874, 2006, pp. 280–290.

[Zsc06b] _____, *Nicht-disjunkte Überdeckungen mit Ferrers-Relationen*, 2006, not published.

[Zsc07a] _____, *Bipartite ferrers-graphs and planar concept lattices*, Proc. of ICFCA07, LNAI 4390, 2007, pp. 313–327.

[Zsc07b] _____, *An FDP-algorithm for drawing lattices*, Proc. of CLA07, 2007, http://www.lirmm.fr/cla07, pp. 62–75.

[Zsc07c] _____, *Relationen mit Ferrers-Dimension 3*, 2007, not published.

[Zsc08] _____, *The number of plane diagrams of a lattice*, Proc. of ICFCA08, LNAI 4933, 2008, pp. 106–123.

Index

bound, 81

component
    m-, 82
    of a Ferrers-graph, 77
    of type 1, 96
    of type 2, 96
concept, 31
connected, 81
connection, 79
consecutive-one property, 48
context, 30
    formal, 30
    standard, 32
corresponding function, 28
crossing number, 27

diagram, 17
    additive, 18
    attribute additive, 19
    layer, 18
    layered attribute additive, 20
    line, 18
    optimal, 27
    plane, 21
    similarity of, 29
dimension, 14
    Ferrers-, 32
    order, 14
    product, 15
dominance drawing, 23

edge crossing, 27

FPC, 44

free triple, 81

graph
    -of an order, 14
    comparability, 14
    Ferrers-, 33

interval
    of a Ferrers-graph, 78
irreducible, 16

lattice, 15
    complete, 15
    concept, 31
    dismantlable, 23
    inclusion-interval, 22
    interordinal, 41
    planar, 21
left-relation
    on contexts, 72
    on diagrams, 58
    on lattices, 36

maximal chain, 28

order, 13
    conjugate, 21
    left-, 39
    linear, 13

poset, 13

relation
    conjugate, 21
    Ferrers-, 32
    incomparability, 13

INDEX

   induced, 72
   induced by a component, 83
   neighborhood, 14
rSPC, 47

sorting relation
   on diagrams, 57
   on lattices, 36
SPC, 46

TPC, 50

Die VDM Verlagsservicegesellschaft sucht für wissenschaftliche Verlage abgeschlossene und herausragende

Dissertationen, Habilitationen, Diplomarbeiten, Master Theses, Magisterarbeiten usw.

für die kostenlose Publikation als Fachbuch.

Sie verfügen über eine Arbeit, die hohen inhaltlichen und formalen Ansprüchen genügt, und haben Interesse an einer honorarvergüteten Publikation?

Dann senden Sie bitte erste Informationen über sich und Ihre Arbeit per Email an *info@vdm-vsg.de*.

Sie erhalten kurzfristig unser Feedback!

VDM Verlagsservicegesellschaft mbH
Dudweiler Landstr. 99
D - 66123 Saarbrücken
www.vdm-vsg.de

Telefon +49 681 3720 174
Fax   +49 681 3720 1749

Die VDM Verlagsservicegesellschaft mbH vertritt

Printed by Books on Demand GmbH, Norderstedt / Germany